PHYSICS FROM THE GROUND UP

**HERMAN Y. CARR,
RICHARD T. WEIDNER**
PROFESSORS OF PHYSICS
RUTGERS UNIVERSITY

PART III — SPECIAL RELATIVITY AND QUANTUM THEORY
— The Physics of the Very Fast and the Very Small

ROBERT E. KRIEGER PUBLISHING COMPANY
HUNTINGTON, NEW YORK
1981

Original Edition 1971 (1 Volume)
Reprint Edition 1981 (Parts 1, 2 & 3) with corrections

Printed and Published by
ROBERT E. KRIEGER PUBLISHING COMPANY, INC.
645 NEW YORK AVENUE
HUNTINGTON, NEW YORK 11743

Copyright © 1971 by
McGRAW-HILL, INC.
Reprinted by Arrangement

All rights reserved. No reproduction in any form of this book, in whole or in part (except for brief quotation in critical articles or reviews), may be made without written authorization from the publisher.

Printed in the United States of America

Library of Congress Cataloging in Publication Data

Carr, Herman Y., 1924-
   Physics from the ground up.

   Reprint of the edition published by McGraw-Hill, New York.
   Includes index.
   1. Physics.   I. Weidner, Richard T., joint author.   II. Title.
[QC21.2.C36 1980]   530   78-22000
ISBN 0-89874-213-7 (Part 3)

TO AMANDA AND ROBERT,
CHRISTOPHER, ALLEGRA, AND TIMOTHY

TO THE READER

There are many reasons for studying the workings of the physical universe; we assume you have your reasons and a knowledge of elementary algebra including some trigonometry. Otherwise, very little is required as a prerequisite to this book.

As the title suggests, our approach to an understanding of physics is through the simple behavior of ordinary objects. We do not seek to accumulate a multitude of results, but instead to trace out a few generalizations of considerable breadth and power. Rather than covering a lot, our strategy is to uncover the most important.

The text falls into five main areas:

- The classical conservation laws—mass, momentum, and energy (Chapters 1 to 11)
- The classical interactions—gravitational, electric, and magnetic (Chapters 12 to 16)
- Light and waves (Chapters 17 to 20)
- The physics of the very fast—relativity (Chapters 21 and 22)
- The physics of the very small—quantum theory, atoms, nuclei, and elementary particles (Chapters 23 to 28)

In developing these topics we have tried to choose arguments that are concise, yet inescapably clear. These conditions are frequently met by the actual historical approach—for example, Christian Huygens' introduction of energy.

A word about some special notations used in this book. The symbol **+** marks a section that is *extra*, in the sense that it is an interesting extension of the main theme but is not essential to the continuing story. The symbol **c** marks other extra items for which *calculus* is required. A reader unfamiliar with calculus may simply skip this material.

Examples worked out and explained in detail are set off from the main body of the text by a shaded background.

For help in preparing "Physics from the Ground Up" the authors are indebted first of all to students and instructors at Rutgers College who used and contributed to the improvement of preliminary versions. We are also indebted to Mr. David A. Beckwith of McGraw-Hill, who contributed suggestions and criticisms in unusual measure. Our special appreciation goes to Professor Robert Ehrlich, who prepared the computer-produced electron-density pictures appearing in Chapter 26. One of us (HYC), whose lecture notes comprised the first version of this book, owes a special debt of gratitude to three teachers: Professors Sergio De Benedetti, Gerald Holton, and Edward M. Purcell.

HERMAN Y. CARR
RICHARD T. WEIDNER

CONTENTS

TO THE READER   v

CHAPTER

1   THE LAW OF INERTIA AND THE CONCEPT OF MASS   2
- 1-1   Objects as Particles   2
- 1-2   When Is an Object Isolated?   3
- 1-3   The Law of Inertia   6
- 1-4   Velocity   8
- + 1-5   The Uniformity of Space   14
- 1-6   Isolated Systems   15
- 1-7   The Concept of Mass   15
- 1-8   Systems of Units   19
- Summary   21
- Problems   21

CHAPTER

2   CONSERVATION LAWS FOR MASS AND FOR MOMENTUM   26
- 2-1   Conservation of Mass   26
- 2-2   Momentum   27
- 2-3   Conservation of Momentum   28
- Summary   34
- Addendum   Momentum Conservation and Reference Frames   35
- Problems   36

CHAPTER

3   FORCE AND THE EXCHANGE OF MOMENTUM   40
- 3-1   The Concept of Force   40
- 3-2   The Units of Force   42
- 3-3   The Superposition Principle for Forces   44
- 3-4   Equilibrium   46
- 3-5   Acceleration   48
- 3-6   Acceleration from the Force of Gravity   52
- 3-7   Mass and Weight   55
- Summary   58
- Problems   59

CHAPTER 4 MORE ON FORCE AND THE EXCHANGE OF MOMENTUM — 64

- 4-1 Motion at Constant Acceleration — 64
- 4-2 Newton's Three Laws of Motion — 70
- 4-3 Force Measurements with a Spring — 72
- Summary — 78
- Problems — 78

CHAPTER 5 MOMENTUM AND FORCE IN GLANCING COLLISIONS — 82

- 5-1 Conservation of Momentum — 83
- 5-2 Vector Components — 87
- 5-3 Force and Acceleration — 94
- 5-4 Superposition of Forces — 98
- 5-5 Equilibrium — 102
- 5-6 Uniform Circular Motion — 103
- Summary — 108
- Problems — 109

CHAPTER 6 CENTER OF MASS AND REFERENCE FRAMES — 114

- + 6-1 Velocity and Position of the Center of Mass — 114
- + 6-2 Acceleration of the Center of Mass — 118
- + 6-3 Frames of Reference — 120
- + 6-4 Inertial Frames — 122
- + 6-5 Invariance of Momentum Conservation — 124
- + 6-6 Collisions in the Center-of-mass Reference Frame — 125
- Summary — 127
- Problems — 127

CHAPTER 7 ANGULAR MOMENTUM — 131

- + 7-1 Introduction to Angular Momentum and Its Conservation — 132
- + 7-2 The Angular Momentum of a Particle — 134
- + 7-3 The Law of Angular-momentum Conservation — 141
- + 7-4 Angular Momentum and Torque for a Single Particle — 144
- + 7-5 Angular-momentum Conservation Derived — 147
- + 7-6 Rotational Equilibrium — 149
- Summary — 151
- Problems — 152

CHAPTER 8 — KINETIC ENERGY AND WORK — 157

- 8-1 The Retrieval of Relative Speed in a Collision — 158
- + 8-2 Coefficient of Restitution and Elastic Collisions — 163
- 8-3 The Retrieval of *Vis Viva* — 166
- 8-4 Kinetic Energy and Work — 167
- 8-5 Units of Kinetic Energy and Work — 170
- + 8-6 Work Done by a Variable Force — 173
- Summary — 176
- Problems — 177

CHAPTER 9 — ENERGY CONSERVATION AND POTENTIAL ENERGY — 181

- 9-1 Energy Conservation and the Concept of Potential Energy — 182
- 9-2 The Potential Energy of a Spring — 185
- 9-3 Gravitational Potential Energy — 189
- + 9-4 Nonisolated Systems and Nonconservative Forces — 193
- Summary — 197
- Problems — 198

CHAPTER 10 — THE KINETIC THEORY OF DILUTE GASES — 203

- 10-1 Properties of Dilute Gases — 204
- 10-2 The Perfect-gas Law (Experiment) — 206
- 10-3 The Kinetic Model of a Dilute Gas — 208
- 10-4 The Ideal-gas Law (Theory) — 210
- 10-5 Avogadro's Hypothesis — 213
- 10-6 Temperature and Kinetic Energy — 215
- Summary — 217
- Problems — 217

CHAPTER 11 — TEMPERATURE, HEAT, AND INTERNAL ENERGY — 219

- 11-1 Thermal Energy — 220
- 11-2 Thermal Equilibrium and Temperature — 222
- 11-3 Heat and Work — 224
- 11-4 The First Law of Thermodynamics — 225
- 11-5 The Second Law of Thermodynamics — 228
- Summary — 231
- Problems — 231

CONTENTS

CHAPTER 12 — THE GRAVITATIONAL INTERACTION — 235

12-1	Gravitational Charge and Gravitational Mass	237
12-2	Kepler's Laws of Planetary Motion	239
12-3	Newton and Gravity	240
12-4	The Moon and the Apple	243
+ 12-5	The Gravitational Interaction of Spherical Objects	246
12-6	The Cavendish Experiment	249
+ 12-7	Gravitational Potential Energy	252
12-8	The Gravitational Field	260
	Summary	261
	Problems	261

CHAPTER 13 — THE ELECTRIC INTERACTION — 265

13-1	The Concept of Electric Charge	266
13-2	Electric-charge Conservation	270
13-3	Electric-charge Quantization	271
13-4	Coulomb's Law of Electric Interaction	271
13-5	The Electric Field	275
+ 13-6	The Electric Field of a Continuous Distribution of Charge	279
+ 13-7	Energy Contained in the Electric Field	284
	Summary	285
	Problems	286

CHAPTER 14 — MORE ON THE ELECTRIC INTERACTION — 289

14-1	Energy Conservation and Electric Potential	290
+ 14-2	Electric Potential Energy for a Radial Interaction Force	294
+ 14-3	Equipotential Surfaces	296
+ 14-4	Conductors	297
	Summary	302
Addendum 1	The Electroscope	302
Addendum 2	The Van de Graaff Generator	303
	Problems	304

CHAPTER 15 — THE MAGNETIC INTERACTION — 309

15-1	Magnets and Moving Charges	310
15-2	The Magnetic Interaction between Two Moving Point Charges	312
15-3	Electric Current, Resistance, and Ohm's Law	313

	15-4	Magnetic Interaction between a Moving Point Charge and a Long Straight Current-carrying Wire	315
	15-5	The Magnetic Field	319
	15-6	The Magnetic Force	322
	15-7	Charged Particles in a Uniform Magnetic Field	323
+	15-8	The Magnetic Force on Current-carrying Conductors	329
+	15-9	The Magnetic Force between Two Parallel Wires	332
+	15-10	The Magnetic Field of a Current Loop	333
+	15-11	Magnetic Force on a Current Loop	336
+	15-12	Magnets	337
		Summary	338
Addendum	+	The Magnetic Force and Reference Frames	339
		Problems	341

CHAPTER
16   CHANGING ELECTRIC AND MAGNETIC FIELDS   345

	16-1	Magnetically Induced Currents	346
	16-2	Electromotance	347
	16-3	Magnetic Flux	353
	16-4	Faraday's Law of Induced Electric Fields; Lenz' Law	357
+	16-5	Induced Electromotance in Moving Conductors	361
+	16-6	Maxwell's Law of Induced Magnetic Fields	366
		Summary	369
		Problems	370

CHAPTER
17   LIGHT AND THE CLASSICAL PARTICLE THEORY   375

17-1	Some Familiar Observations	376
17-2	The Classical Particle Theory	378
17-3	Some Predictions of the Particle Model	379
17-4	The Speed of Light	380
17-5	Snell's Law	383
17-6	Refraction and the Particle Theory	388
17-7	Newton's Rings and Young's Experiment	390
	Summary	393
	Problems	393

CHAPTER
18   THE WAVE THEORY AND INTERFERENCE OF LIGHT   397

	18-1	Wave Propagation along a Line of Coupled Masses	398
+	18-2	Mathematical Derivation of the Wave Speed	400

	18-3	Other Types of Waves	404
	18-4	Superposition of Waves	405
	18-5	Reflection of Waves	407
	18-6	Periodic Waves	410
	18-7	The Double-slit Experiment	411
	18-8	Color and Vision	414
+	18-9	The Grating Spectrometer	416
+	18-10	Newton's Rings	420
	18-11	Status Report on the Wave and Particle Theories of Light	422
		Summary	423
		Problems	424

CHAPTER 19 DIFFRACTION 427

	19-1	Propagation of Waves as a Beam	428
	19-2	Wavefronts and Huygens' Principle	432
	19-3	Reflection and Refraction of Waves	433
	19-4	Fresnel and Fraunhofer Diffraction at a Single Slit	437
		Summary	444
		Problems	445

CHAPTER 20 ELECTROMAGNETIC WAVES 447

	20-1	The Concept of an Electromagnetic Wave	448
	20-2	The Electromagnetic Spectrum	452
+	20-3	Emission and Absorption of Electromagnetic Waves	454
+	20-4	Radiation Force and the Momentum of Light	460
		Summary	465
		Problems	465

CHAPTER 21 TIME DILATION AND SPACE CONTRACTION 467

	21-1	The Speed of Light: A Universal Constant?	468
	21-2	Reference Frames and the Velocity of Light	470
	21-3	Time Dilation	474
	21-4	Length Contraction	480
+	21-5	The Relativity of Simultaneity	487
	21-6	Relativistic Velocity Transformations	489
+	21-7	The Relativity of Electric and Magnetic Fields	495
		Summary	497
Addendum	+	The Michelson-Morley Experiment	497
		Problems	501

CHAPTER 22 RELATIVISTIC DYNAMICS — 505

- 22-1 Momentum and the Relativity of Mass — 506
- 22-2 Relativistic Momentum — 511
- 22-3 Relativistic Energy — 512
- 22-4 Rest Mass and Energy — 516
- 22-5 Particles of Zero Rest Mass — 518
- 22-6 Units and Computations in Relativistic Dynamics — 521
- Summary — 524
- Problems — 524

CHAPTER 23 PARTICLE PROPERTIES OF WAVES — 527

- 23-1 Experimental Description of the Photoelectric Effect — 528
- 23-2 The Inadequacy of Classical Concepts — 530
- 23-3 The Quantum Theory of the Photoelectric Effect — 532
- 23-4 Photons as Particles — 536
- + 23-5 The Compton Effect — 537
- Summary — 543
- Problems — 544

CHAPTER 24 ENERGY QUANTIZATION IN ATOMIC SYSTEMS: THE BOHR MODEL OF THE HYDROGEN ATOM — 547 / 548

- 24-1 The Nuclear Atom — 549
- 24-2 The Inadequacy of a Classical Planetary Atomic Model — 552
- 24-3 The Bohr Model of the Hydrogen Atom — 558
- + 24-4 The Correspondence Principle — 562
- + 24-5 Quantization of Angular Momentum — 564
- Summary — 565
- Problems

CHAPTER 25 WAVE PROPERTIES OF PARTICLES — 567

- 25-1 De Broglie's Prediction of Matter Waves — 568
- + 25-2 Observation of the Wavelengths of X-radiation — 570
- 25-3 The Observation of Matter Waves — 575
- 25-4 The Heisenberg Uncertainty Principle — 575
- + 25-5 Principle of Complementarity — 585
- 25-6 The Probability Interpretation of Matter Waves — 586
- Summary — 588
- Problems — 589

CHAPTER 26 QUANTUM SYSTEMS OF PARTICLES — 591

- 26-1 Waves along a String — 592
- 26-2 Waves over a Circular Membrane — 596
- 26-3 The Puddle and the Particle in a Saucer — 600
- 26-4 The Wave Theory of the Hydrogen Atom — 603
- \+ 26-5 Electron Spin and the Four Quantum Numbers — 607
- \+ 26-6 The Periodic Table — 610
- \+ 26-7 Groups of Atoms — 613
- Summary — 620
- Problems — 621

CHAPTER 27 NUCLEAR PHYSICS — 623

- \+ 27-1 The Proton-electron Nuclear Model — 624
- \+ 27-2 Discovery of the Neutron — 625
- 27-3 The Nuclear Force — 629
- 27-4 Nuclear Binding Energy — 630
- 27-5 The Conservation Laws in Radioactive Decay — 634
- \+ 27-6 The Radioactive-decay Law — 638
- \+ 27-7 Nuclear Reactions — 644
- Summary — 647
- Problems — 648

CHAPTER 28 ELEMENTARY-PARTICLE PHYSICS — 651

- 28-1 The Universal Conservation Laws — 652
- 28-2 Pair Production and Pair Annihilation — 655
- 28-3 Conservation of Electron Number — 659
- 28-4 Conservation of Baryon Number — 662
- \+ 28-5 Beta Decay Revisited — 667
- \+ 28-6 Conservation of Muon Number — 671
- \+ 28-7 The Fundamental Interactions — 674
- \+ 28-8 The Nonconservation of Parity in Weak Interactions — 679
- Summary — 685
- Problems — 686

ANSWERS TO SELECTED PROBLEMS — 689

INDEX — 697

TIME DILATION AND SPACE CONTRACTION

CHAPTER TWENTY-ONE

CHAPTER 21
TIME DILATION AND SPACE CONTRACTION

The basic concepts of physics—momentum, force, energy, etc.—rest finally on measurements of space and time. When we speak of a force, for example, we mean the *rate in time* at which an object's momentum changes. Momentum is mass times velocity, and velocity is obtained as the quotient of the *space* interval between two events and the corresponding *time* interval. Even the determination of mass reduces, in principle, to the measurement of speeds after a collision and therefore to measurements of space and time.

Physics, then, might not unjustly be called the study of space and time intervals. From an operational point of view we can all agree on how these intervals are to be defined. A space interval, or length, is what is measured by applying a meterstick; a time interval is what is measured by counting the ticks of a regular clock. But will all observers get the same number for the same interval? According to common sense, and to our tacit assumption in developing Newtonian mechanics, there is no doubt about it: they will. Still, we have never really investigated the question, and the paradox on page 464 may indicate that something is wrong with the Newtonian picture at very high speeds. In the present chapter, we shall find that a coherent description of physical events is quite impossible unless everyday notions of space and time are drastically revised.

21-1 THE SPEED OF LIGHT: A UNIVERSAL CONSTANT?

The central result of Chapter 20 was the deduction that an electromagnetic wave travels through a vacuum at the constant speed $c = \sqrt{k_e/k_m} = 3.0 \times 10^8$ m/s. This speed is determined solely by the fundamental electric and magnetic constants, whose values can be found by measuring the interaction forces between stationary or moving charges in vacuum. As far as experiments show, the *laws* governing the interaction of charged bodies are universal, independent of the state of motion of the observer. To see what this implies, imagine a man in a fast train and a man in the station, both of whom observe the same two charged particles. They will, of course, obtain different values for the velocities of the particles and for the interaction force between them. In fact, one man might say that that force is purely electric, while the other finds it both electric and magnetic. However, both of them can agree on the amounts of charge carried by the particles, and both will find *the same formulas* connecting force, charge, velocity, and separation. In particular, both observers will attribute the same value to k_e and the same value to k_m.

This universal nature of electric and magnetic interaction has a profound consequence: if k_e and k_m are indeed universal constants, so is the speed of light. All observers, however they may move with respect to each other, should find that an electromagnetic disturbance travels through empty space at 3×10^8 m/s!

Universal constancy of the speed of light predicted from electricity and magnetism

This conclusion seems *fantastic* because it is so contradictory to our understanding of wave motion. Newton, and later Maxwell (who first derived $c = \sqrt{k_e/k_m}$), believed that a subtle yet material "ether" permeated all space. If light were a disturbance of this ether, then, as we believe on the basis of Chapter 18, the speed of a light wave through the ether would be determined by that medium's intrinsic properties. Observers at rest with respect to the ether would measure the intrinsic speed, but observers moving relative to the ether would measure a different speed, depending on their own motion. For example, if the velocity of an observer relative to the ether were 0.4 times the intrinsic speed c and in the same direction as the light wave, he would measure a wave speed of $0.6c$; but going the opposite way, he would measure $1.4c$. Indeed, he would always be able to tell whether he was at rest in the ether by checking whether the measured wave speed was precisely c. Such conclusions are easy to accept because they are perfectly analogous to the more familiar waves, e.g., sound waves, which propagate by means of a material medium. There was little reason to question these conclusions until Maxwell's work provided the basis for the contradictory conclusion that *the speed of light past an observer is the universal constant* $c = \sqrt{k_e/k_m}$, regardless of how the observer is moving.

Universal constancy of the speed of light seems to be contradictory to the wave theory of light, at least for waves propagating in a material medium.

Experiment, the way the universe behaves, must be the final test of all theories, hypotheses, and assumptions. Is it actually possible to detect a difference between two precise measurements of the speed of light made by observers in relative motion? The difficulties are formidable. First of all, the sought-for difference will be very small unless the relative speed of the two observers is comparable to the speed of light itself. About the best one can do is to utilize the revolution of the Earth about the sun, and this gives a relative speed of only 6×10^4 m/s, or less than one-thousandth the speed of light. Second, it is hard enough to measure c with accuracy, let alone to measure small variations in c. Clearly, any experiment which attempts to check on the constancy of c must be of remarkable ingenuity and precision. Nevertheless, just such an experiment, one of the most elegant of the nineteenth century, was performed in 1887 by Michelson and Morley (see the Addendum to this chapter). The accuracy of their measurements was more than adequate to allow detection of a speed difference of the expected size, but no such difference was found—by them or by subsequent experimenters.

The universal constancy of the speed of light is a possibly perplexing, but nonetheless firmly established, fact of nature. We shall use it as the starting point for the special theory of relativity, developed by Albert Einstein (1879–1955) in 1905. Einstein actually based his own confidence in the invariance of c solely on considerations in electricity and magnetism. The theory of relativity was epochal both in scientific and general cultural thought, much as the Copernican revolution had been before it. In expanding our conceptions of space and time, it affected the entire structure of physics.

21-2 REFERENCE FRAMES AND THE VELOCITY OF LIGHT

The universal constancy of the speed of light cannot be reconciled with our usual rule for connecting the different velocities of a disturbance (wave or particle) measured by observers in different reference systems. Something has to give. To see what that might be, let us review the derivation of this rule, this time keeping a sharp eye out for "self-evident" but actually untested assumptions.

Once again it will be helpful to speak in terms of two observers, one in a railroad station and one in a train passing with constant speed V, who view the same phenomenon, in this case the propagation of a wave pulse.

The station (Fig. 21-1) has a roof supported by three posts, one at each end and one in the middle. A reference frame R, the xyz coordinate system, is at rest with respect to the station; a second reference frame R', the $x'y'z'$ system, is at rest with respect to the train which moves to the right with speed V relative to the station and frame R. A long Slinky coil is strung from the middle post to the post at the right end of the platform, and a rope is strung from the midpoint of the train across the last seat of the front car, shown in Fig. 21-1.

Just as the midpoint of the train passes the middle post (Fig. 21-1b) a special event occurs: the left end of the Slinky is momentarily displaced. As a result a pulselike disturbance, the top of which is visible to the observer on the train, travels to the right (Fig. 21-1c).

Classical velocity transformation derived

It is easy to write an equation relating the velocity v_x of this disturbance as measured in R to the velocity v'_x of the same disturbance as measured in R'. First take the point of view of the observer in the station. For him the pulse begins at time t_1 and reaches the righthand post at the t_2, that is, at the end of a time interval $\Delta t =$

471

SECTION 21-2
REFERENCE
FRAMES AND
THE VELOCITY
OF LIGHT

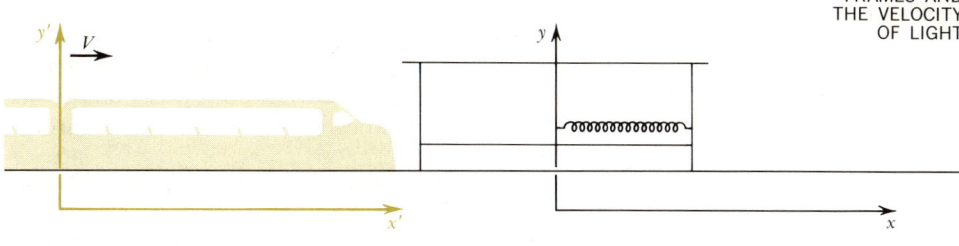

(a) $t < t_1$

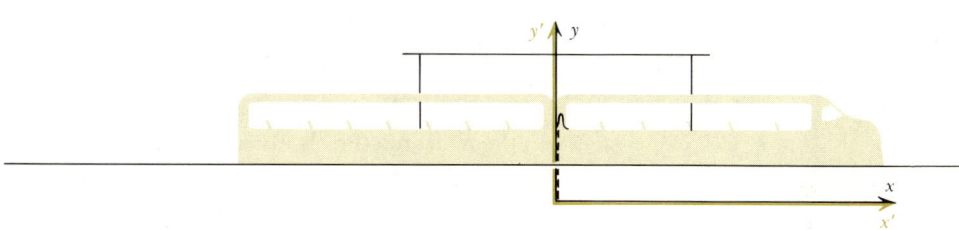

(b) $t = t_1$

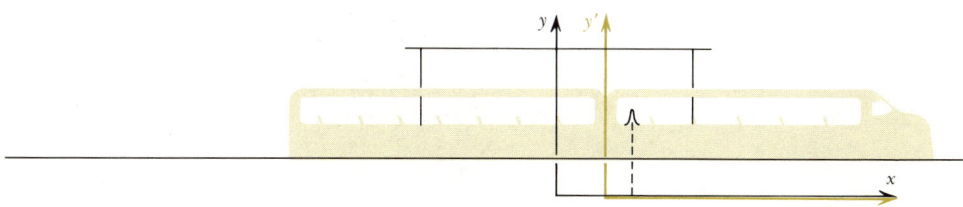

(c) $t_1 < t < t_2$

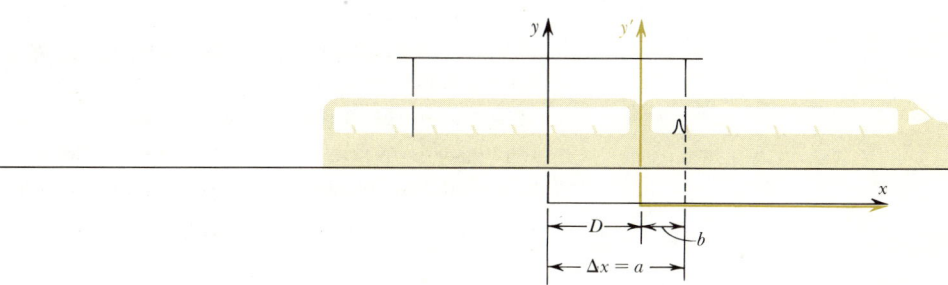

(d) $t = t_2 = t_1 + \Delta t$

FIGURE 21-1 (a) A train (reference frame R') moves to the right relative to the station (reference frame R) at speed V. (b) A pulse begins at $t = t_1$ at the left end of the Slinky and (c) travels to the right. (d) At $t = t_2$ the pulse arrives at the right post.

$t_2 - t_1$. During this same time interval the midpoint of the train travels a distance D to the right which falls short of the right post by just the length of the rope strung across the last seat of the front car. This second event, the end of the disturbance at time t_2, is shown in Fig. 21-1d.

In reference frame R, the net displacement of the pulse during the time interval Δt is $\Delta x = x_2 - x_1 = a$, where a is the length of the coil and x_1 and x_2 are the coordinates of the two posts between which the coil is stretched. Thus the velocity of the Slinky disturbance as calculated in R is

$$[21\text{-}1] \quad v_x = \frac{\Delta x}{\Delta t} = \frac{a}{\Delta t}$$

In the R' frame the observer sees the first event (the start of the Slinky disturbance) occur at a location x'_1, which is the midpoint of the train. The second event, the end of the disturbance, occurs at a point x'_2 farther forward in the train, at the front end of the rope strung across the last seat in the first car (see Fig. 21-2). The magnitude of this displacement is just $\Delta x' = x'_2 - x'_1 = b$, where b is the length of the rope. Thus the R' observer finds the velocity of the Slinky disturbance to be

$$[21\text{-}2] \quad v'_x = \frac{\Delta x'}{\Delta t} = \frac{b}{\Delta t}$$

But wait! Why have we written Δt, and not $\Delta t'$, as the time interval in R' between the start and end of the Slinky disturbance? Because we are *assuming* that all observers, no matter what their reference frame, measure the same time interval between the same two events. "My time is your time" is so successful a principle in everyday life that we have never questioned whether it is always true. As we go on, we must remember that the conclusions which follow are based on this assumption.

Implicit but untested assumption that all observers measure the same time interval between the same two events

It is easy to relate v'_x to v_x; we simply look at Figs. 21-1 and 21-2 and find the following two relationships, respectively:

[21-3] In reference frame R: $a = b + D$

[21-4] In reference frame R': $a = b + D$

Implicit but untested assumption that all observers measure the same length for the same object

As you probably already suspect, we have written separately what appear to be identical equations in order to emphasize a crucial point. Indeed, [21-4] should be written $a' = b' + D'$. The only basis for writing it just like [21-3] is another previously successful assumption which we have never questioned before: we have always assumed

SECTION 21-2
REFERENCE
FRAMES AND
THE VELOCITY
OF LIGHT

(a) $t' < t'_1$

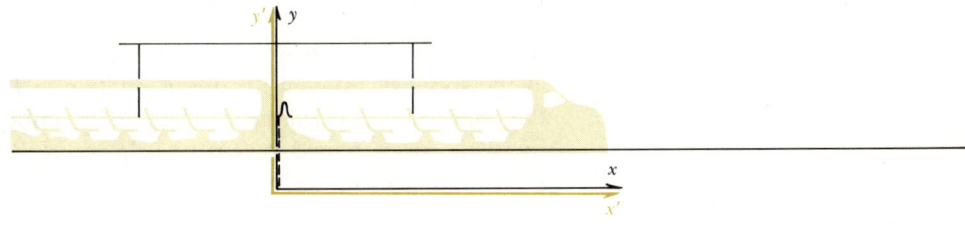

(b) $t' = t'_1$

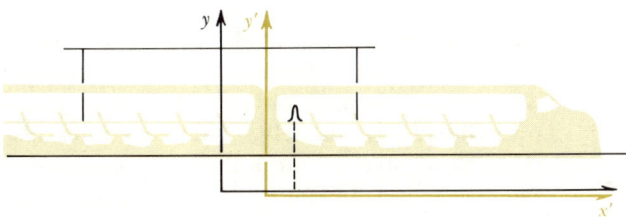

(c) $t'_1 < t' < t'_2$

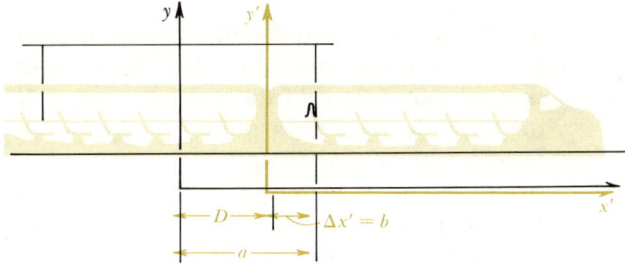

(d) $t' = t'_2$

FIGURE 21-2 The events of Fig. 21-1 as viewed by an observer (R') at rest with respect to the train and for whom the station is in motion at speed V. Again a disturbance travels along the Slinky from one end to the other.

that all observers, no matter what their velocity relative to an object is, will measure the same length for that object so long as they mark the positions of the two ends simultaneously. Therefore the conclusions which follow are also based on this assumption.

To obtain the relationship between v_x and v'_x we simply divide both sides of [21-3] (or [21-4]) by the time interval Δt between the beginning and end of the Slinky disturbance, and using [21-1] and [21-2], we immediately find

[21-5] $$v_x = v'_x + V$$

where V, the speed of reference frame R' relative to R, is the same as the speed of reference frame R relative to R'. Equation [21-5] is the usual rule for combining velocities. It simply says that the velocity of the disturbance in R is its velocity in R' plus the velocity of R' with respect to R.

For the motion of ordinary objects and of material waves [21-5], if not exact, is a superb approximation; for light (to which the foregoing reasoning should apply unchanged), it is wildly wrong.

The essence of the scientific approach since Galileo has been that whenever an assumption, however comfortable, comes into collision with a fact, the fact prevails. Two suppositions have been seen to underlie [21-5], namely, that time intervals and space intervals (lengths of objects) are universal. One or both of these must now be abandoned. In the next two sections we shall find the relations that hold in their stead.

21-3 TIME DILATION

The possibility that different observers measure different time intervals between the same two events can be explored by modifying the thought experiment of the previous section. This time the observer in the station sets off a pulse of light (event 1), using a flashbulb fixed to the central post. The pulse travels vertically upward and strikes (event 2) a mirror mounted on the ceiling of the station. Reflected straight back down, it returns (event 3) to the bulb (see Fig. 21-3). Our observer in R uses his meterstick and clock to measure the vertical distance H between flashbulb and mirror and the time interval Δt between events 1 and 3. He thus determines the speed of light as

[21-6] $$c = \frac{2H}{\Delta t}$$

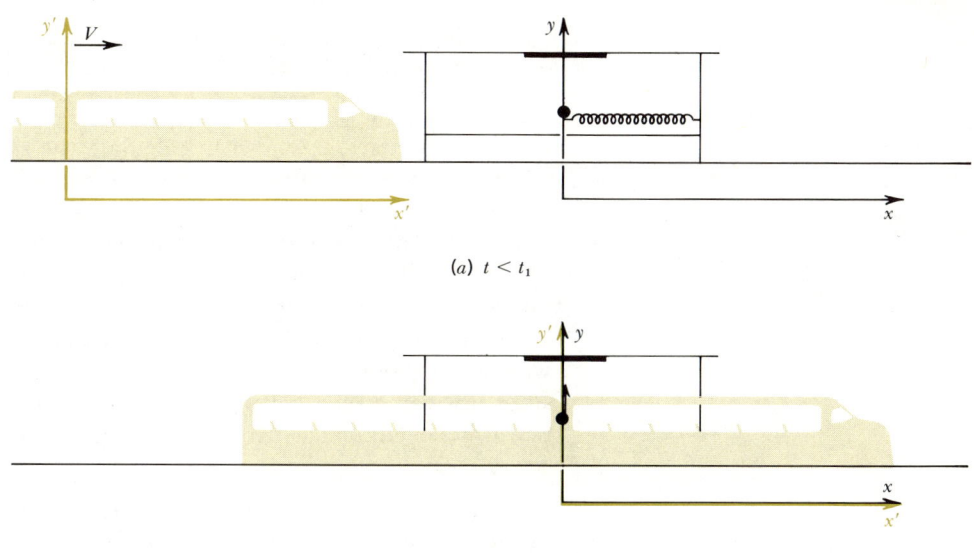

(a) $t < t_1$

(b) t_1

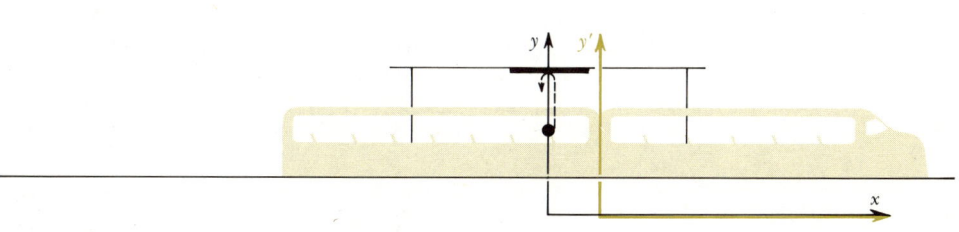

(c) t_2

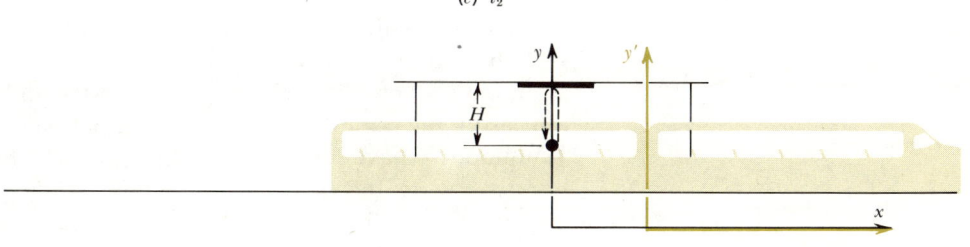

(d) $t_3 = t_1 + \Delta t$

FIGURE 21-3 An observer in the station sets off a light pulse at t_1 which travels up, strikes a mirror on the ceiling at t_2, and arrives back at the point of origin at t_3.

Notice that, in contrast to the Slinky experiment, the initial and terminal events occur, for the R observer, *at the same place.*

What about the R' observer on the train? He too can use measurements of events 1, 2, and 3 to calculate the speed of light, and he will obtain the same number, $c = 3 \times 10^8$ m/s. In R' events 1 and 3 have *different* locations, because the bulb moves a distance $\Delta x'$ during the time of flight $\Delta t'$ of the light pulse. (Figure 21-4 shows the light path as observed in R'.) To assure a perfectly definite time measurement we can replace the single R' observer by a whole group of observers, provided with stopwatches synchronized beforehand, who sit all along the train. An observer at the middle of the train stops his watch when event 1 occurs, and an observer toward the rear times event 3. The two men may then compare readings to find $\Delta t'$.

If the relative speed between train and station is V, and if H' is the distance from bulb to mirror as measured in R', then

$$\Delta x' = V \Delta t'$$

and, by applying the Pythagorean theorem in R',

$$c(\tfrac{1}{2}\Delta t') = \sqrt{(H')^2 + (\tfrac{1}{2}\Delta x')^2}$$

or

[21-7]
$$c = \frac{2\sqrt{(\tfrac{1}{2}\Delta x')^2 + (H')^2}}{\Delta t'} = \frac{\sqrt{(V\Delta t')^2 + (2H')^2}}{\Delta t'}$$

It is now possible to eliminate the vertical distances H and H' from [21-6] and [21-7]. The crucial observation is this: although we must question whether observers moving with different velocities will measure the same length for a length parallel to the line of relative motion, we have less reason to question whether lengths measured transverse to the line of the motion are the same. After all, the longitudinal velocities measured for the object are different, but the transverse velocities are all the same. Therefore we take $H = H'$ in the two expressions for c, [21-6] and [21-7] and obtain

$$c^2 = V^2 + c^2 \left(\frac{\Delta t}{\Delta t'}\right)^2$$

or

[21-8]
$$\Delta t' = \frac{\Delta t}{\sqrt{1 - (V/c)^2}}$$

SECTION 21-3
TIME
DILATION

(a) $t' < t'_1$

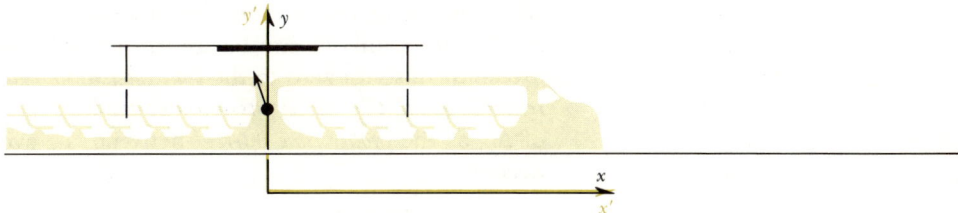

(b) t'_1

(c) t'_2

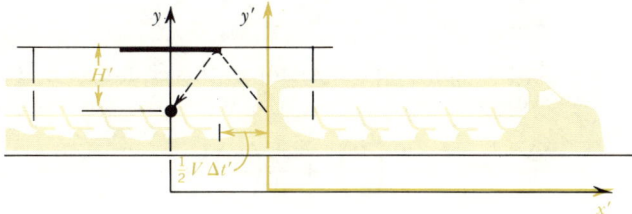

(d) $t'_3 = t'_1 + \Delta t'$

FIGURE 21-4 The events of Fig. 21-3, as viewed by an observer (R') traveling with the train. For this observer the pulse leaves at t'_1, travels obliquely upward and strikes the mirror at t'_2, and arrives at the bulb at t'_3.

CHAPTER 21
TIME DILATION AND SPACE CONTRACTION

The universal constancy of the speed of light implies that observers in different reference frames do not measure the same time interval between the same two events.

The two time intervals, $\Delta t'$ measured by observers seated on the train and Δt measured by observers standing on the station platform, are not the same; $\Delta t'$ is longer than Δt.

Does this mean that observers on a moving train always measure times that are longer than those measured by observers on a station platform? Not at all. We must remember that the result [21-8] has been deduced for a special set of events. The departure of the light from the bulb and the subsequent arrival of the light back at the bulb occurred at the same position in the reference frame at rest with respect to the station but at different positions in the frame fixed in the train. To emphasize this underlying asymmetry we shall attach a subscript zero to the *time interval measured by observers in whose reference frame the two events occur at the same position*. We shall speak of Δt_0 as the *proper time interval* between the events. Equation [21-8] becomes

[21-9] $$\Delta t' = \frac{\Delta t_0}{\sqrt{1 - (V/c)^2}}$$

Observers in the reference frame for which the two events occur at the same position measure the shortest time interval.

from which it is seen that the proper time interval is the shortest time interval elapsing between a pair of events. The interval measured by an observer in any reference frame moving at constant velocity relative to the proper reference frame will be larger than Δt_0. For him the proper interval will have been dilated; hence the name *time dilation*.

We can explore the meaning of [21-9] further by finding its form for low speeds:

$$\frac{\Delta t'}{\Delta t_0} = \frac{1}{\sqrt{1 - (V/c)^2}} = \left[1 - \frac{V^2}{c^2}\right]^{-1/2}$$

Expanding the last factor in this equation by the binomial theorem, we have

$$\frac{\Delta t'}{\Delta t_0} = 1 + \frac{1}{2}\left(\frac{V}{c}\right)^2 + \frac{3}{8}\left(\frac{V}{c}\right)^4 + \cdots$$

If $V \ll c$, we can discard the term in $(V/c)^4$, so that [21-9] gives

[21-10] $$\frac{\Delta t'}{\Delta t_0} \approx 1 + \frac{1}{2}\left(\frac{V}{c}\right)^2$$

To a first approximation, then, $\Delta t' = \Delta t_0$, which corresponds to the normal assumption that time intervals are the same for all observers. As far as everyday occurrences are concerned, this assumption is a very good one. Even a difference in intervals of only $\frac{1}{2}$ percent requires a speed ratio satisfying

$$\frac{\frac{1}{2}}{100} = \frac{1}{2}\left(\frac{V}{c}\right)^2$$

or

$$V = \tfrac{1}{10}c \approx 18{,}600 \text{ miles/s}$$

Small wonder that time dilation goes unnoticed; human observers simply don't move relative to each other with sufficiently high speeds.

Nevertheless, time dilation has been observed directly. The lifetimes of certain unstable particles (muons) associated with cosmic rays from outer space are measured as they are moving toward the observers on Earth at speeds very nearly equal to the speed of light. These lifetimes are found to be longer than those of muons which are at rest with respect to the Earth, just as demanded by [21-9].

The stretching of time with motion is no less startling than the constancy of the speed of light, but the one follows from the other. Once more we see the danger in believing that our immediate perceptions tell the whole story about the universe. Einstein may not have been too harsh when he defined common sense as "that layer of prejudices laid down in the mind prior to the age of eighteen."

EXAMPLE 21-1

The distance from point A over New York City to a point B over Miami, Florida, is 1,000 miles as measured by observers at rest with respect to the Earth. A Martian rocket travels between these two points in 8.3×10^{-3} s as measured by observers at rest with respect to the Earth. How long do the observers on the rocket, who use their own clocks, determine that the trip takes? (For convenience take $c = 2 \times 10^5$ miles/s.)

SOLUTION

The two events, the rocket at point A over New York and the rocket at point B over Miami, occur at the same position in a frame of reference $x'y'z'$ at rest with respect to the rocket. Thus the dilated interval $\Delta t = 8.3 \times 10^{-3}$ s measured by the observers on Earth is related to the proper interval $\Delta t'_0$ measured by the observers on the rocket by

$$\Delta t = \frac{\Delta t'_0}{\sqrt{1 - (V/c)^2}}$$

where V, the speed of the rocket relative to the Earth, is given by

$$V = \frac{\Delta x}{\Delta t} = \frac{1{,}000 \text{ miles}}{8.3 \times 10^{-3} \text{ s}} = 1.2 \times 10^5 \text{ miles/s}$$

Therefore

$$\Delta t'_0 = \Delta t \sqrt{1 - \left(\frac{V}{c}\right)^2} = 8.3 \times 10^{-3} \sqrt{1 - \left(\frac{1.2 \times 10^5}{2 \times 10^5}\right)^2}$$

$$= 6.6 \times 10^{-3} \text{ s}$$

EXAMPLE 21-2 The time interval between the production (birth) of a particle and its decay (death) is called its *lifetime*. The lifetime of a certain particle is 4.0×10^{-6} s as measured in the system at rest with respect to the particle. The particle moves at a speed $0.6c$ relative to the laboratory. What is the lifetime of the particle as measured by observers at rest with respect to the laboratory?

SOLUTION The two events (birth and death) occur at the same place in the frame of reference at rest with respect to the particle. Let us *arbitrarily* call this the reference frame *xyz*. The proper time interval measured in this frame is $\Delta t_0 = 4.0 \times 10^{-6}$ s. The time interval measured in the moving laboratory reference frame $x'y'z'$ will be dilated, and is given by

$$\Delta t' = \frac{\Delta t_0}{\sqrt{1 - (V/c)^2}} = \frac{4.0 \times 10^{-6}}{\sqrt{1 - 0.6^2}} = 5.0 \times 10^{-6} \text{ s}$$

21-4 LENGTH CONTRACTION

Another railroad thought experiment will reveal that not only time intervals but space intervals as well, i.e., lengths, vary with the observer.

So that the newfound relationship [21-9] will apply, the experiment is again designed with the initial and final events occurring at the same position in one reference frame, the frame R fixed in the station. However, we now want to consider a light signal moving *parallel to* the line of motion of the train, which can be accomplished by having a pulse from the flashbulb on the middle post reflected back to the source by a mirror mounted vertically on the right-hand post. The three events of interest are, as previously, (1) the departure of the light pulse from the bulb, (2) its arrival at the mirror, and (3) its subsequent arrival back at the bulb (see Fig. 21-5). As measured in R, the time interval between events 1 and 3 is the proper time interval Δt_0. We have

SECTION 21-4
LENGTH
CONTRACTION

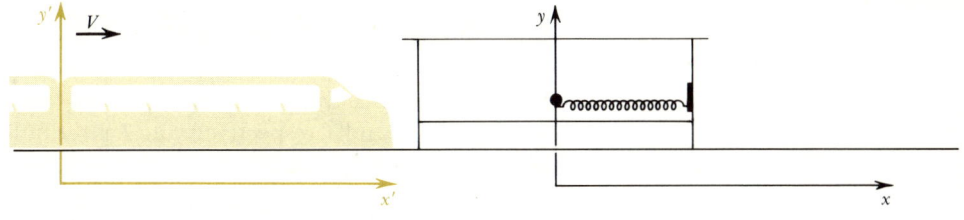

(a) $t < t_1$

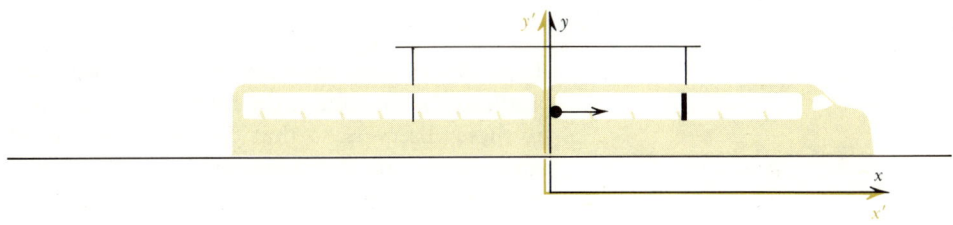

(b) t_1

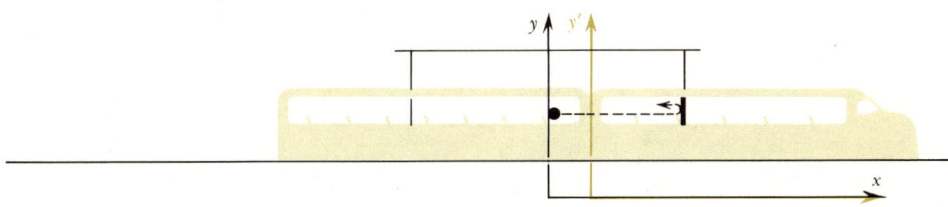

(c) t_2

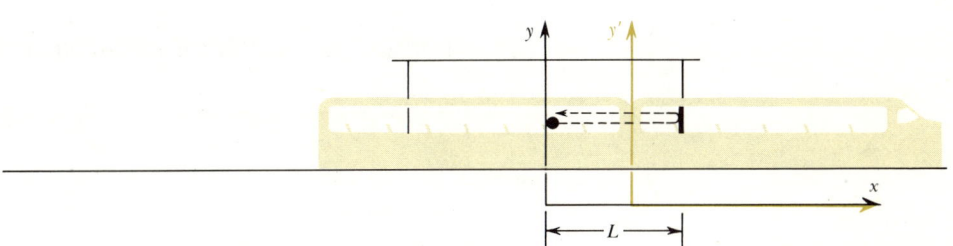

(d) $t_3 = t_1 + \Delta t_0$

FIGURE 21-5 A light signal is sent parallel to the line of motion of the train. The light pulse leaves the bulb at t_1, reaches the mirror at t_2, and returns to the bulb at t_3.

[21-11] $$\Delta t_0 = \Delta t_{12} + \Delta t_{23} = \frac{L}{c} + \frac{L}{c} = \frac{2L}{c}$$

Here, Δt_{12} and Δt_{23} are the time intervals between events 1 and 2 and between 2 and 3, respectively and L is the bulb-to-mirror distance, all as determined in R.

For observers in R' (on the train) the three events in question have the locations shown in Fig. 21-6. Using their clocks, they obtain (in analogy to [21-11])

[21-12] $$\Delta t' = \Delta t'_{12} + \Delta t'_{23}$$

In addition, they measure the spatial separation between the bulb and mirror. Of course, the bulb and mirror are moving with respect to these observers, so that they must be careful to make *simultaneous* observations of the positions of the two objects in order to get a correct measurement of their separation. Let us suppose they have managed this and the result is L'. Then, by the constancy of the speed of light, we have (see Fig. 21-6),

[21-13] $$c\,\Delta t'_{12} = L' - V\,\Delta t'_{12} \qquad c\,\Delta t'_{23} = L' + V\,\Delta t'_{23}$$

from which

[21-14] $$\Delta t'_{12} = \frac{L'}{c+V} \qquad \Delta t'_{23} = \frac{L'}{c-V}$$

Substituting [21-14] into [21-12] gives

[21-15] $$\Delta t' = \frac{L'}{c+V} + \frac{L'}{c-V} = L'\frac{2c}{c^2 - V^2} = \frac{2L'/c}{1 - (V/c)^2}$$

But $\Delta t'$, as given by [21-15], and the proper interval Δt_0, as given by [21-11], are connected by the time-dilation formula [21-9]. Thus,

$$\frac{2L'/c}{1 - (V/c)^2} = \frac{2L/c}{\sqrt{1 - (V/c)^2}}$$

or

[21-16] $$L' = L\sqrt{1 - (V/c)^2}$$

Our conclusion is that space intervals (lengths), like time intervals, do depend on the frame of reference of the observer. However, the transformation of space intervals is reciprocal in form to time intervals. According to [21-16], the observer on the train, who is in motion with respect to the end points of the interval, measures a *smaller* length than the observer in the station does. Equation [21-16] is therefore said to describe *length contraction*.

SECTION 21-4
LENGTH
CONTRACTION

(a) $t' < t'_1$

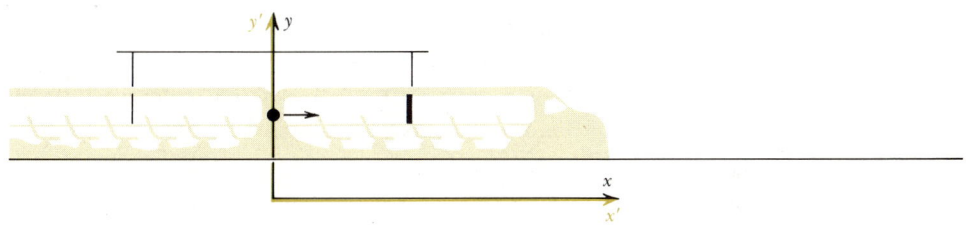

(b) t'_1

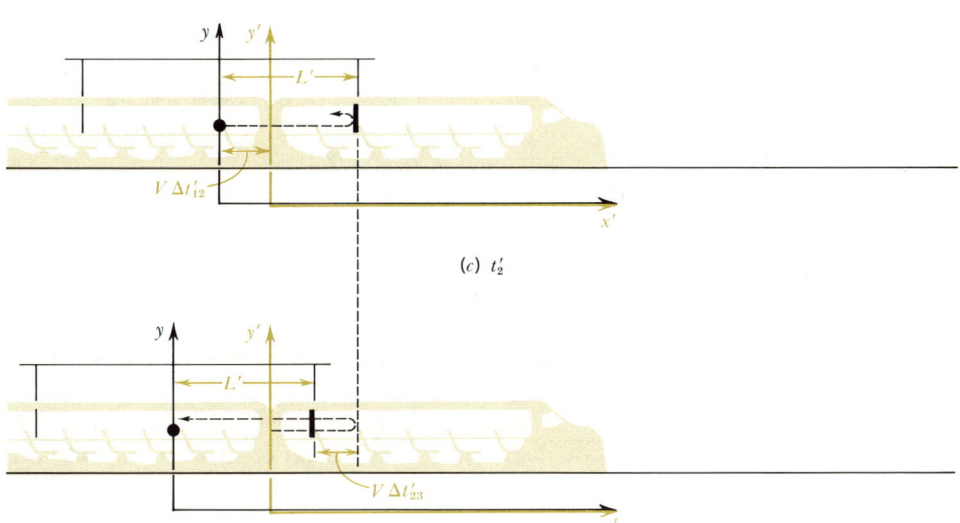

(c) t'_2

(d) $t'_3 = t'_1 + \Delta t'$

FIGURE 21-6 The events of Fig. 21-5 as viewed by an observer (R') traveling with the train. For this observer the pulse leaves the bulb at t'_1, reaches the mirror at t'_2, and returns to the bulb at t'_3.

CHAPTER 21
TIME DILATION AND SPACE CONTRACTION

The universal constancy of the speed of light implies that observers in different reference frames do not measure the same length for the same object.

Observers in the reference frame for which the object is at rest measure the longest length.

[21-17]

Everything observed about time dilation applies to length contraction, provided we interchange the words "longer" and "shorter." Thus, the *proper length* L_0 of a space interval or material rod (the length in a reference frame at rest with respect to the end points) is its longest length. Any observer in uniform motion with respect to the interval will determine a shorter length L', according to

$$L' = L_0\sqrt{1 - (V/c)^2}$$

It should be emphasized that the contraction of a material body which accompanies its motion is not like the contraction which results from, say, a lowering of its temperature. The effect originates from the basic properties of space. As we have already seen, the contraction effect is restricted to lengths along the line of relative motion; transverse dimensions are the same for both observers.

EXAMPLE 21-3 Imagine that the flashbulb and mirror of Fig. 21-5 are located on the train rather than on the station platform. A light pulse is (1) emitted from a flash bulb at the rear of a car, (2) reflected from a mirror at the front end of the car, and (3) detected back at the bulb. If observers at rest with respect to the train measure a distance L_0' from the bulb to the mirror, what distance do observers at rest with respect to the station measure for this distance? Assume that all observers carry identical clocks and all know that light signals have a constant speed c.

SOLUTION For observers on the station platform (Fig. 21-7), $\Delta t = \Delta t_{12} + \Delta t_{23}$, where $c\,\Delta t_{12} = L + V\,\Delta t_{12}$, $c\,\Delta t_{23} = L - V\,\Delta t_{23}$, and L is the length from rear to front of the car. Thus

$$\Delta t = \Delta t_{12} + \Delta t_{23} = \frac{L}{c-V} + \frac{L}{c+V} = \frac{2Lc}{c^2 - V^2} = \frac{2L/c}{1 - (V/c)^2}$$

For observers on the train (Fig. 21-8), $\Delta t_0' = 2L_0'/c$, and because the two events 1 and 3 occur at the same place relative to the train, we know from time dilation that

$$\Delta t = \frac{\Delta t_0'}{\sqrt{1-(V/c)^2}} = \frac{2L_0'/c}{\sqrt{1-(V/c)^2}}$$

Thus equating our two expressions for Δt, we have

$$L = L_0'\sqrt{1 - \left(\frac{V}{c}\right)^2}$$

In this example, it is the distance measured by the observers on the platform that is contracted.

SECTION 21-4
LENGTH
CONTRACTION

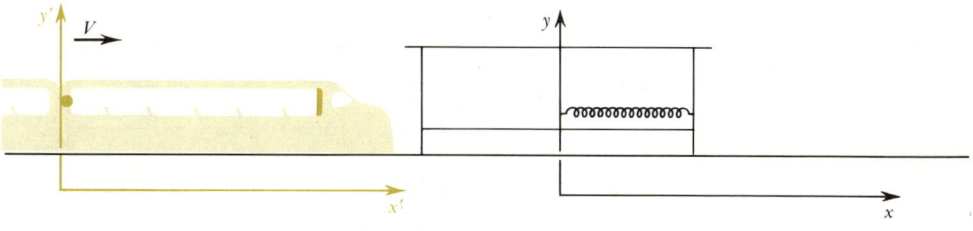

(a) $t < t_1$

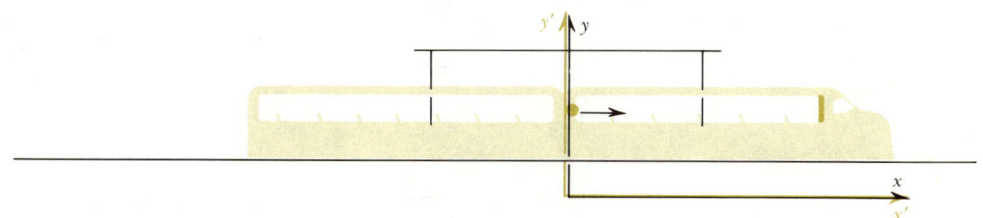

(b) t_1

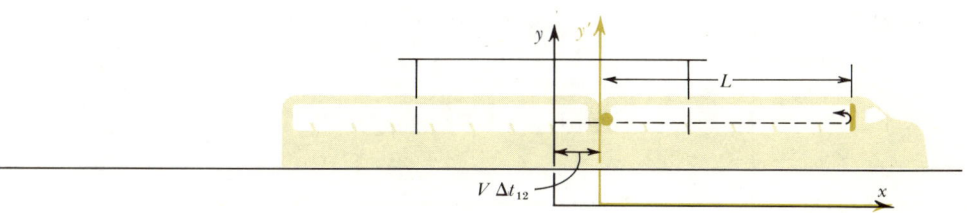

(c) t_2

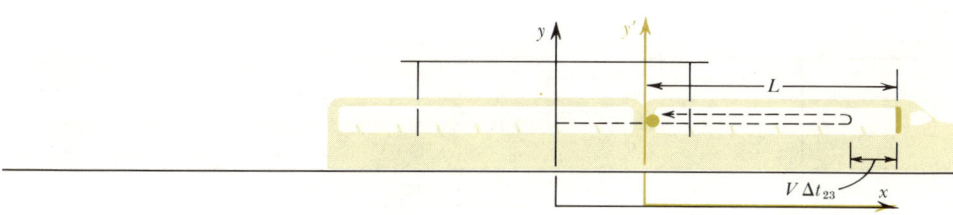

(d) $t_3 = t_1 + \Delta t$

FIGURE 21-7

CHAPTER 21
TIME DILATION AND SPACE CONTRACTION

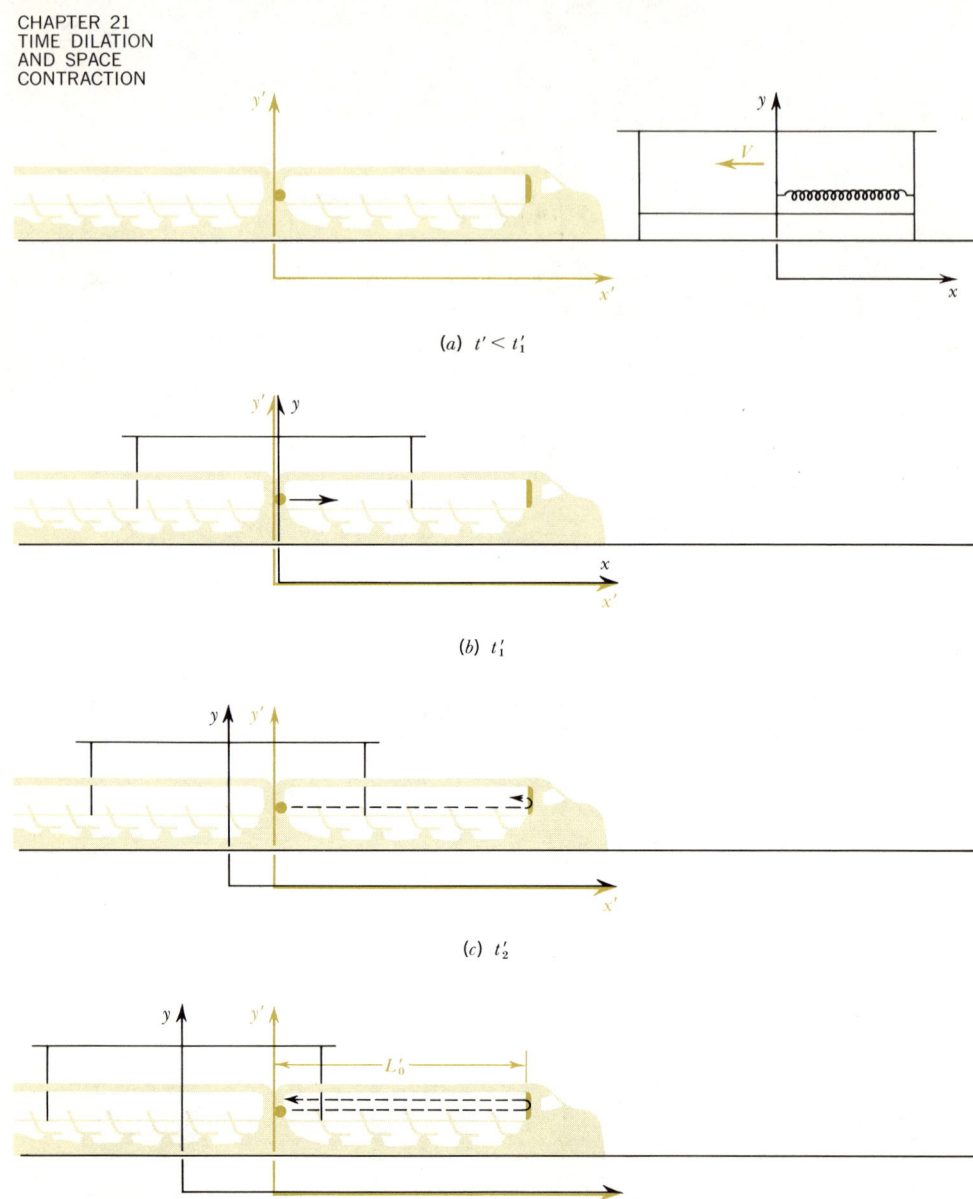

(a) $t' < t'_1$

(b) t'_1

(c) t'_2

(d) $t'_3 = t'_1 + \Delta t'_0$

FIGURE 21-8

EXAMPLE 21-4

In Example 21-1, how far do observers *on the rocket* find that point B over Miami is from point A over New York?

SOLUTION

If observers at rest with respect to Miami and New York measure a separation of 1,000 miles between these two points, observers moving with speed V relative to these points and along the line connecting them will measure a contracted distance of $1,000\sqrt{1-(V/c)^2}$ miles. From Example 21-1, $V = 1.2 \times 10^5$ miles/s. Hence, the contracted length is $1,000\sqrt{1-(1.2/2)^2} = 800$ miles.

Check: According to the observers on the rocket ship it takes 6.6×10^{-3} s (see Example 21-1) for this 800-mile interval on Earth to pass the ship. It follows that the speed of the Earth relative to the ship is 800 miles/$(6.6 \times 10^{-3}$ s$) = 1.2 \times 10^5$ miles/s, which is identical, as it should be, to the value obtained in Example 21-1 for the speed of the ship relative to the Earth.

21-5 THE RELATIVITY OF SIMULTANEITY

The very concept of time is closely bound up with the notion of simultaneous events. One of Einstein's profoundest insights was that events which are simultaneous in one reference frame need not be so in another. In fact, if the two events are spatially separate in the frame in which they are simultaneous, they will generally occur at different times in another frame moving relative to the first! To see how this *relativity of simultaneity* follows from the constancy of the speed of light, back to the station.

Imagine that there are flashbulbs attached to both the left- and right-hand posts of the railway station, as shown in Fig. 21-9. If an observer standing on the platform at the middle post receives two light signals, one from each end of the platform, at the same time (event 3), he knows that one signal was emitted from the left end of the platform (event 2) at the same time that the other signal was emitted from the right end of the platform (event 1). After all, he is at the middle of the platform, the bulbs are at the ends, and light signals always move with the same speed c. In his reference frame R, then, events 1 and 2 are simultaneous.

The observers riding in the train are at rest with respect to reference frame R' moving at constant speed V relative to R. They see event 3 corresponding to the two light signals arriving together at the middle post of the station and call this time t'_3.* As for events 1 and 2, they note that the light signal from the left post moves in the opposite direction to the motion of the station, while the signal

---

*Event 3 is not "split" in R'; for since the proper time interval between the two arrivals is zero, any dilated interval is also zero. More basically, there is no relativity of events: if an event (everything that happens at one place at one time) occurs in one reference frame, it occurs in every reference frame.

488

CHAPTER 21
TIME DILATION
AND SPACE
CONTRACTION

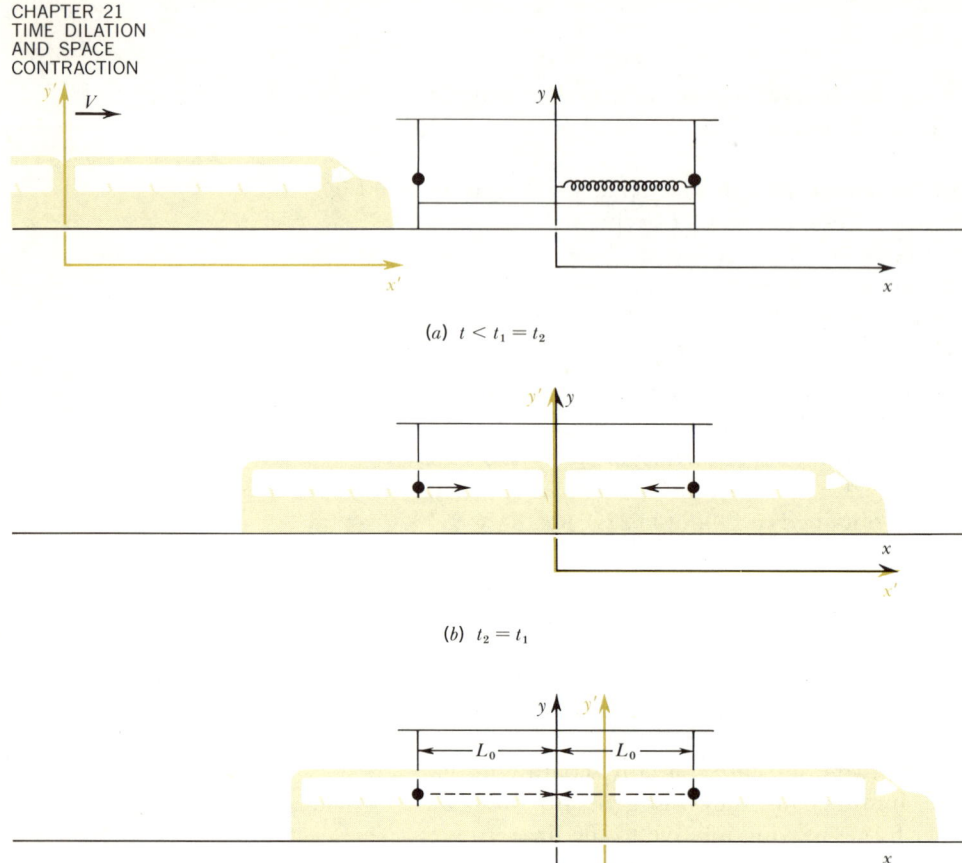

(a) $t < t_1 = t_2$

(b) $t_2 = t_1$

(c) t_3

FIGURE 21-9 An observer standing on a platform at its middle point receives two light signals, one from each end, which arrive simultaneously at his location.

If observers in one reference frame find that two events occurring at different positions are simultaneous, then observers in other reference frames will find that these same two events are not simultaneous.

from the right post moves in the same direction as the station. Thus the light signal from the left post, with a shorter distance to travel to the meeting point, must have been emitted later in time than the other signal, it being known that the speed of both signals is the same. In other words, in the train frame of reference the two events 1 and 2 are *not* simultaneous.

Having deduced the relativity of simultaneity, we can go further and determine the actual difference in time $t'_1 - t'_2$ in this situation. Because the station is moving to the left with speed V relative to the train, observers at rest with respect to the train find that the light signal directed from the left post of the station toward the middle post travels a distance $V \Delta t'_{23}$ less than the distance L' between posts (see Fig. 21-10). The signal from the right post of the station travels a distance $V \Delta t'_{13}$ greater than this length L'. Thus,

$$c \Delta t'_{23} = L' - V \Delta t'_{23} \quad \text{and} \quad c \Delta t'_{13} = L' + V \Delta t'_{13}$$

or

$$t'_2 - t'_1 = \Delta t'_{13} - \Delta t'_{23} = \frac{L'}{c - V} - \frac{L'}{c + V} = \frac{2L'V/c^2}{1 - (V/c)^2} \qquad [21\text{-}18]$$

We must remember that the station is in motion relative to the train; consequently, observers at rest with respect to the train measure the contracted length $L' = L_0 \sqrt{1 - (V/c)^2}$, where L_0 is the distance between posts measured by observers standing on the station platform. Thus,

$$t'_2 - t'_1 = \frac{2L_0 V/c^2}{\sqrt{1 - (V/c)^2}} = \frac{(x_1 - x_2)V/c^2}{\sqrt{1 - (V/c)^2}} \qquad [21\text{-}19]$$

where $x_1 - x_2$ is the spatial separation of the two events which are simultaneous as observed in reference frame R at rest with respect to the station. Notice that for no spatial separation ($x_1 = x_2$), the two events are always simultaneous ($t'_1 = t'_2$).

21-6 RELATIVISTIC VELOCITY TRANSFORMATIONS

In Section 21-1 we discovered that the usual rule $v'_x = v_x - V$ does not correctly express the relationship between velocities in different reference frames. The error was traced to two commonsense assumptions, which have subsequently been replaced by the concepts of time dilation and length contraction. Now we can proceed to find the right form of the velocity-transformation law.

The experiment to be used is that of Fig. 21-1, slightly modified. A disturbance in the Slinky is, as before, initiated at the center post (event 1), but as soon as the wave reaches the right-hand post, a flash bulb there is ignited (event 2) and a light signal is received back at the middle post (event 3) (see Fig. 21-11). The change permits the application of the time-dilation formula, since it puts events 1 and 3 in the same location in one of the reference frames.

According to observers standing on the station platform, the total time required for the sequence of events is

$$\Delta t_0 = \Delta t_{12} + \Delta t_{23} = \frac{L_0}{v_x} + \frac{L_0}{c} \qquad [21\text{-}20]$$

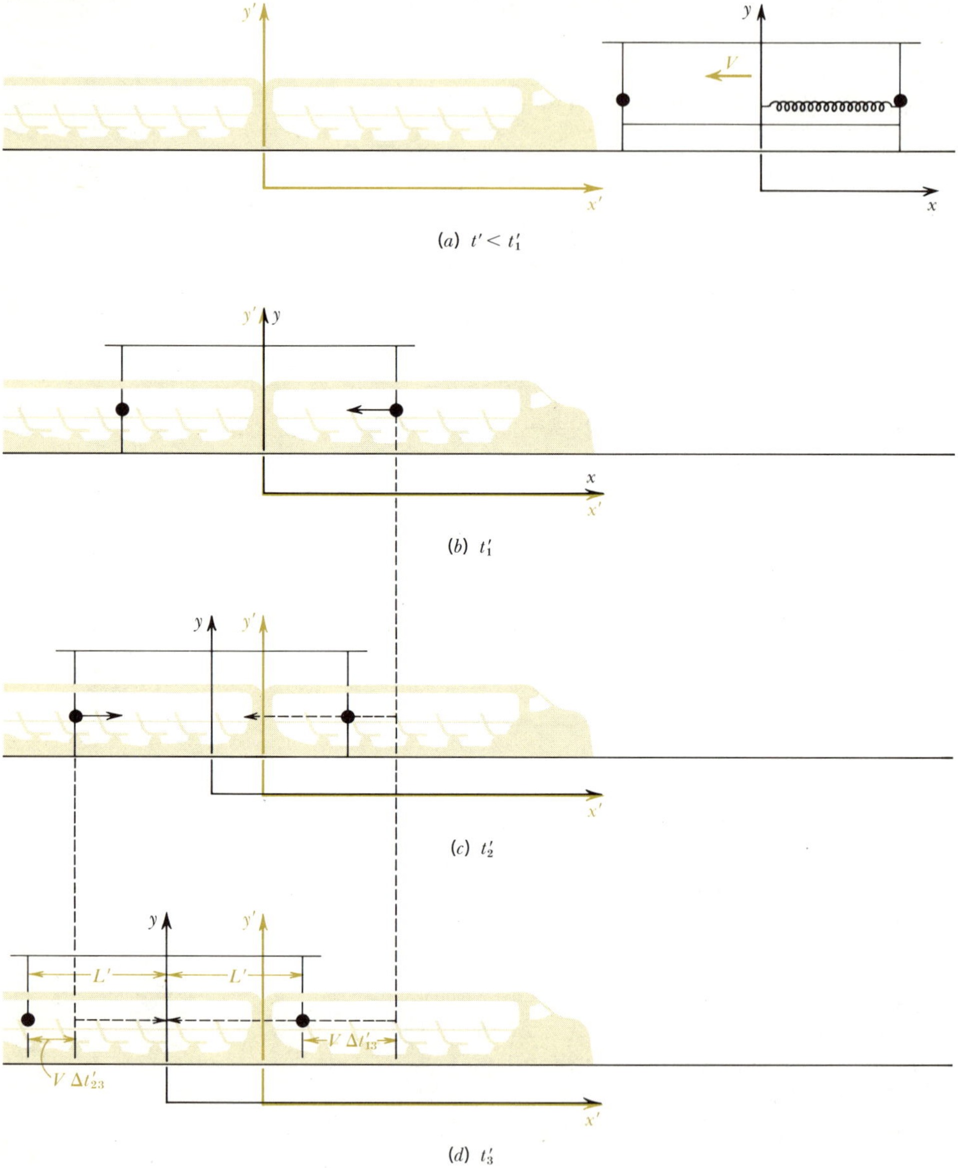

CHAPTER 21
TIME DILATION
AND SPACE
CONTRACTION

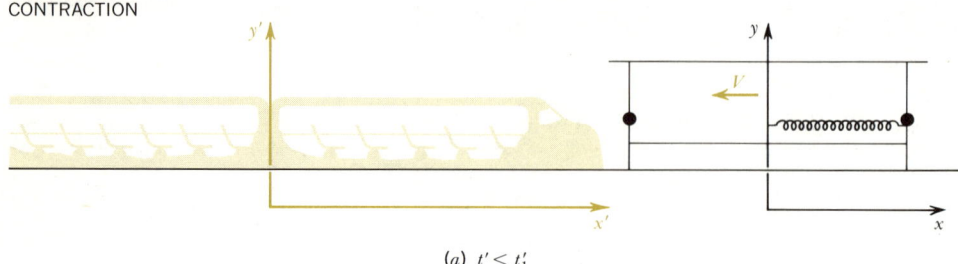

FIGURE 21-10 The events of Fig. 21-9 as observed by someone at rest on the train; relative to this observer the arrivals but not the departures of the two light signals are simultaneous.

491

SECTION 21-6
RELATIVISTIC
VELOCITY
TRANSFORMATIONS

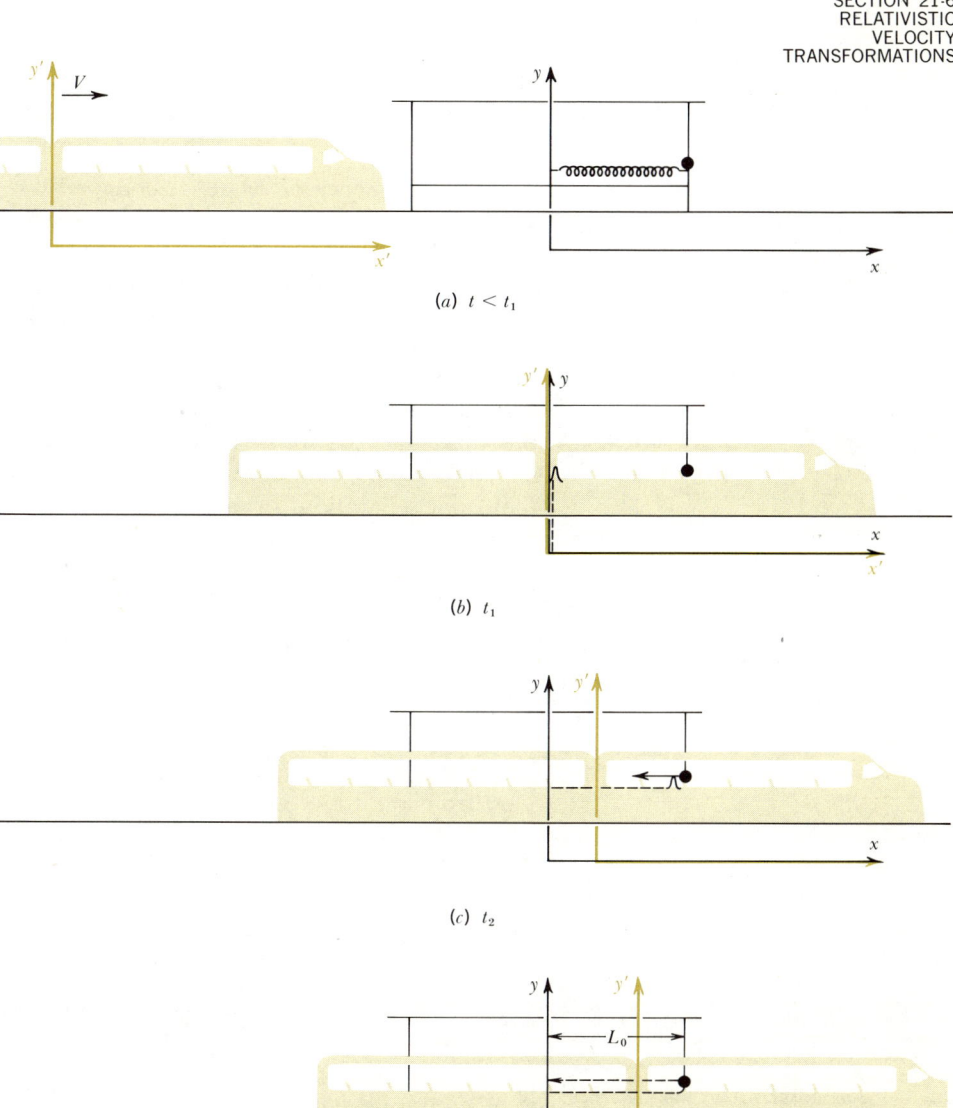

(a) $t < t_1$

(b) t_1

(c) t_2

(d) $t_3 = t_1 + \Delta t_0$

FIGURE 21-11 A pulse is initiated on the Slinky at the center post at t_1. When it arrives at t_2 at the far post, it initiates a light signal, which reaches the center post at t_3.

Here again L_0 is the distance between posts as measured by observers standing on the station platform, and v_x is the speed they measure for the disturbance in the Slinky.

The same sequence of events as observed by persons seated on the train is shown in Fig. 21-12. These observers see the station moving toward the rear of the train at speed V. They also see the Slinky pulse moving toward the front of the train with speed v'_x. This disturbance then collides with the right post of the station, whereupon a light signal starts to move toward the rear of the train. Because the station and the Slinky pulse move in opposite directions, whereas the station and the light signal move in the same direction, they find that the slinky pulse travels a distance less than the observed post separation and the light signal travels a distance greater than that separation. In fact,

Distance slinky pulse travels: $v'_x \Delta t'_{12} = L' - V \Delta t'_{12}$

Distance light pulse travels: $c \Delta t'_{23} = L' + V \Delta t'_{23}$

Solving for the two time intervals, we have

$$\Delta t'_{12} = \frac{L'}{v'_x + V} \qquad \Delta t'_{23} = \frac{L'}{c - V}$$

Therefore, the total time interval between events 1 and 3, as measured from the train, is

[21-21] $$\Delta t' = \Delta t'_{12} + \Delta t'_{23} = L'\left(\frac{1}{v'_x + V} + \frac{1}{c - V}\right)$$

The rest is just algebra. First, Δt_0 and $\Delta t'$, as determined by [21-20] and [21-21], are linked by the time-dilation formula [21-9]. Hence

[21-22] $$L'\left(\frac{1}{v'_x + V} + \frac{1}{c - V}\right) = \frac{L_0}{\sqrt{1 - (V/c)^2}}\left(\frac{1}{v_x} + \frac{1}{c}\right)$$

The relativistic velocity transformation replaces the classical transformation.

Now we use the length-contraction formula [21-17] to eliminate the post separation from [21-22]. The result, after some algebra, can be written

[21-23] $$v'_x = \frac{v_x - V}{1 - v_x V/c^2}$$

which is the relativistic velocity law we were looking for.

According to [21-23], when an object whose velocity is v_x in frame R is viewed from a frame R' which moves parallel to the object and with speed V with respect to R, the object's velocity in R' is not

SECTION 21-6
RELATIVISTIC
VELOCITY
TRANSFORMATIONS

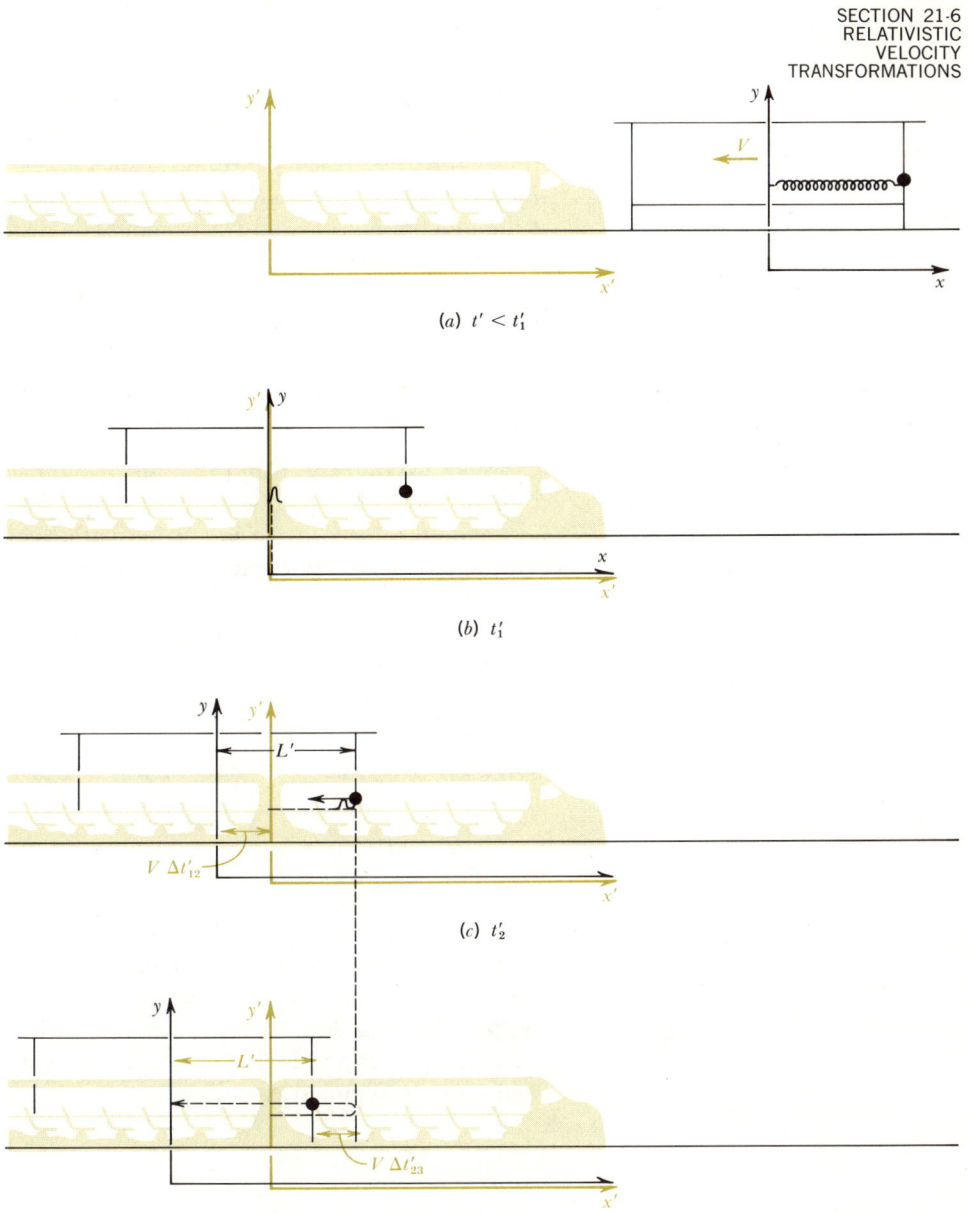

(a) $t' < t'_1$

(b) t'_1

(c) t'_2

(d) $t'_3 = t'_1 + \Delta t'$

FIGURE 21-12 The events of Fig. 21-11 as observed by someone at rest in the train. The Slinky pulse leaves at t'_1 and reaches the far point at t'_2, and the light pulse it triggers there arrives back at the starting point in R' at t'_3.

The relativistic velocity transformation is correct both for very small speeds and for the speed of light.

$v_x - V$ but is larger or smaller, depending on the size of the factor $1/(1 - v_x V/c^2)$. However, when $v_x \ll c$, the factor is very nearly unity, and the classical formula $v'_x = v_x - V$ results. (If this were not the case, we would know that [21-23] was wrong.) Furthermore, if the moving object is an electromagnetic disturbance ($v_x = c$), our new transformation gives

$$v'_x = \frac{c - V}{1 - cV/c^2} = c$$

That is, the speed of light in R' is the same as in R. This, too, must be: the constancy of the speed of light was used to *derive* [21-23].

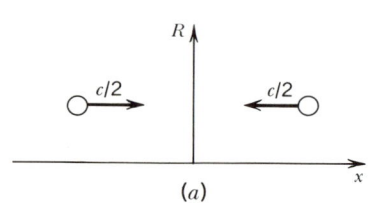

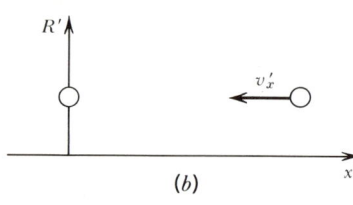

FIGURE 21-13

Although we do expect on the basis of symmetry that the speed of R' relative to R is the same as the speed of R relative to R', we have not made this assumption in our derivations of time dilation, space contraction, or our new velocity transformation [21-23]. This equality of inverse relative speeds is in agreement with [21-23], where we find that for $v_x = 0$, we obtain $v'_x = -V$. But we can now use [21-23] to determine the speed of R' relative to R: we simply solve for v_x when $v'_x = 0$ and obtain $v_x = V$, thus confirming what we expect on the basis of symmetry.

As a final check, we note that when [21-23] is solved for v_x, it takes the form

[21-24] $$v_x = \frac{v'_x + V}{1 + v'_x V/c^2}$$

This says that the velocity v_x in R is calculated from the velocity v'_x in R' and from the relative velocity $-V$ of R with respect to R' *in the same way* as v'_x is calculated from v_x and from V. Put otherwise, we get the same law of transformation whether we regard R as stationary and R' as moving or vice versa. A state of relative motion is an entirely symmetrical affair.

EXAMPLE 21-5   A physicist observes two protons approaching him from opposite directions, each with a speed $0.5c$ relative to his laboratory. At what speed does one proton move relative to the other?

SOLUTION   In the laboratory frame R, the particle on the right side of Fig. 21-13a has velocity $v_x = -c/2$, and the particle on the left side has velocity $+c/2$. But in a frame of reference R' ($V = +c/2$) at rest with respect to the proton on the left, the other proton has a velocity v'_x (see Fig. 21-13b) given by [21-23]

$$v'_x = \frac{v_x - V}{1 - v_x V/c^2} = \frac{-c/2 - (+c/2)}{1 - [(-c/2)(+c/2)]/c^2} = \frac{-c}{1 + \frac{1}{4}} = -\frac{4}{5}c$$

Thus one proton has a speed $0.8c$, not c, relative to the other proton.

The discovery of the universal constancy of the speed of light provided the basis for the special theory of relativity. Because the first indication of this universal constancy appeared in the laws of electricity and magnetism, it will come as no great surprise that simple experiments in electricity and magnetism, experiments which have long been familiar, reveal relativistic effects. What *is* surprising, however, is that the speeds involved in these experiments need not exceed even one-billionth, let alone one-tenth, the speed of light. As a result any dilation of time or contraction of length is exceedingly small. Yet these effects play a crucial role, as the following example will show.

A long straight wire, consisting of equal amounts of positive and negative charge, carries a current of 1 A. In a typical conductor the positive atomic ions remain at rest while the relatively free electrons drift along the wire. Example 15-1 showed that for a 1-A current in a copper wire 1 mm in diameter, the drift velocity is only 0.1 mm/s.

Suppose that a positively charged particle moves parallel to the wire, at the drift speed of the electrons composing the current; the positive charge maintains a radial distance d from the wire (Fig. 21-14). Along every segment of the current-carrying wire there are equal amounts of positive and negative charge. Consequently there is no net electric force exerted on the outside charge. That charge does, however, experience a magnetic force in the direction shown in the figure (if necessary refer back to Figs. 15-11 and 15-15).

Now let us switch over to a new frame of reference at rest with respect to both the charged particle and the free electrons in the wire. What is the force on the charged particle in this frame? Surely, this force, the product of the particle's mass and acceleration, has almost exactly the same direction and magnitude as the force measured in the laboratory system, for the two reference frames have an *extremely* small relative speed. But the interaction force in the new reference frame is a purely electric force: the basic and universal law of magnetic interaction tells us that in a given reference frame there is no magnetic force on a particle at rest in that frame. What then is responsible for this electric force in the new frame of reference? The

+ 21-7 THE RELATIVITY OF ELECTRIC AND MAGNETIC FIELDS

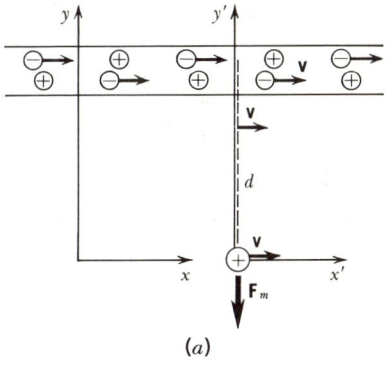

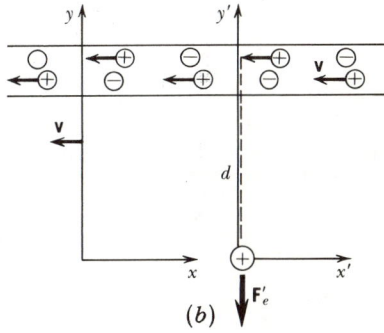

FIGURE 21-14 (*a*) A positively charged particle moves parallel to a wire at the drift speed of the electrons in the wire. A magnetic force F_m acts on this charge. (*b*) The situation shown in (*a*) as viewed by an observer fixed with respect to the positively charged particle. The magnetic force has vanished, but the charge is subject to an electric force F'_e arising from the electric charge unbalance on the wire.

answer is *length contraction*, the minute but definite contraction accompanying a relative speed of only 0.1 mm/s. For suppose the laboratory observers measure a positive charge of $+\lambda$ C in each meter of the wire. If it is assumed that each positive ion has the same amount of charge in any reference frame, observers in the moving frame, where the positive ions are no longer at rest, will find this same amount of charge $+\lambda$ in *less* than 1 m of length. By the same argument, the density of negative charge along the wire has a greater magnitude in the laboratory frame, where the electrons drift, than in the moving frame, in which they are at rest. Therefore, in the moving frame there is more positive charge and less negative charge per meter of wire than in the laboratory frame. Because the two kinds of charge just balance in the laboratory frame, there must be a net positive charge density in the moving frame. This charge distribution gives rise to an electric field directed radially outward from the wire, which is just the direction of the observed force on the outside positive charge.

Even the magnitude of the observed force is exactly accounted for by the relativistic charge imbalance. In terms of the drift speed v (which is the relative speed of the two reference frames) it is shown in Problem 21-19 that the density of positive charge along the wire in the moving frame has the value $\lambda(v/c)^2$ (which is only about $10^{-24}\lambda$). The corresponding electric force (see Problem 13-17) in the moving frame is given by

$$F'_e \approx k_e \frac{2q\lambda(v/c)^2}{d}$$

The magnetic force on a moving charge derived from the electric force observed in a reference frame where the charge is at rest

On the other hand, the magnetic force observed in the laboratory frame of reference (see [15-9] and [15-12]) is

$$F_m = k_m \frac{2v^2\lambda q}{d}$$

Since $k_e/k_m = c^2$, we immediately recognize *that F_m is nothing but the expression of F'_e in the laboratory system.* At last we have the fundamental explanation of magnetism foreshadowed in the Addendum to Chapter 15. The special theory of relativity has shown us that there is really just one kind of interaction between charges, that described by Coulomb's law in the rest frame. The magnetic interaction is just an alias for this basic *electric interaction* in reference frames in which the charged particles are moving.

SUMMARY

If the time interval between two events is Δt_0 when measured by observers at rest with respect to the special (proper) frame of reference *in which the two events occur at the same location,* then observers at rest with respect to another reference frame which moves at constant speed V relative to the special frame measure the dilated time interval

$$\Delta t' = \frac{\Delta t_0}{\sqrt{1 - (V/c)^2}} \qquad [21\text{-}9]$$

If the length of an object is L_0 when measured by observers at rest with respect to the special (proper) reference frame *in which the object is at rest,* then observers at rest with respect to another reference frame which moves at constant speed V relative to the special frame measure the contracted length

$$L' = L_0 \sqrt{1 - \left(\frac{V}{c}\right)^2} \qquad [21\text{-}17]$$

If two events are observed to occur simultaneously in a reference frame in which they occur at separate locations, these two events will not be simultaneous when observed in other reference frames moving with constant speed V relative to the first frame.

If an object is observed to be moving with velocity v_x in one reference frame, its velocity v'_x in a second reference frame moving with constant speed V in the direction of the $+x$ axis is

$$v'_x = \frac{v_x - V}{1 - v_x V/c^2} \qquad [21\text{-}23]$$

+ ADDENDUM
THE MICHELSON-MORLEY EXPERIMENT

Maxwell's electromagnetic theory of light and the presumed universality of the fundamental constants k_e and k_m suggested that the speed of light past an observer is itself a universal constant, regardless of how the observer is moving. It is important that this suggestion be checked by experiment, especially because it is in direct opposition to our usual understanding of wave motion. Any experiment sufficiently sensitive to detect a *change* in the observed speed of light waves resulting from the motion of the observer is very difficult. For example, imagine that a light pulse is transmitted by some medium, or "ether," at rest relative to reference frame R. If our laboratory happens to be at rest with respect to this medium and frame R, the time required for a light signal to move some distance L, for example, the length of a laboratory table, is $\Delta t_1 = L/c$, where c is the intrinsic wave speed of the medium. But if our laboratory is at rest with respect to a reference frame R' which moves in the same direction as the light signal but with the constant speed V relative to frame R and the ether, then the time required for the light signal to travel the length L of the laboratory table is, according to our usual understanding of waves, $\Delta t_2 = L/(c - V)$.

Just how large are these time intervals? The length of a laboratory table is typically about 3 m; thus $\Delta t_1 = 3/(3 \times 10^8)$ or about 10^{-8} s. This is a very short time and difficult to measure to even one significant figure. But the number we are most interested in is the difference between Δt_1 and Δt_2. If we were fortunate enough to have V, the speed of the laboratory relative to the ether, equal to about half the speed of light, this difference would also be the order of 10^{-8} s, but unfortunately the largest V we can reasonably hope to have for our laboratory is the speed with which the Earth orbits the sun, about 3×10^4 m/s. Thus the largest value of the ratio V/c is about 10^{-4}, and the corresponding value of Δt_2 is

$$\Delta t_2 = \frac{L}{c - V} = \frac{L}{c}\frac{1}{1 - V/c} = \frac{L}{c}\left(1 + \frac{V}{c} + \cdots\right) \approx 10^{-8}(1 + 10^{-4})$$

Here we have used the fact that $V/c \ll 1$. The largest possible difference between Δt_1 and Δt_2, a mere 10^{-12} s, is so small as to be impossible to measure directly.

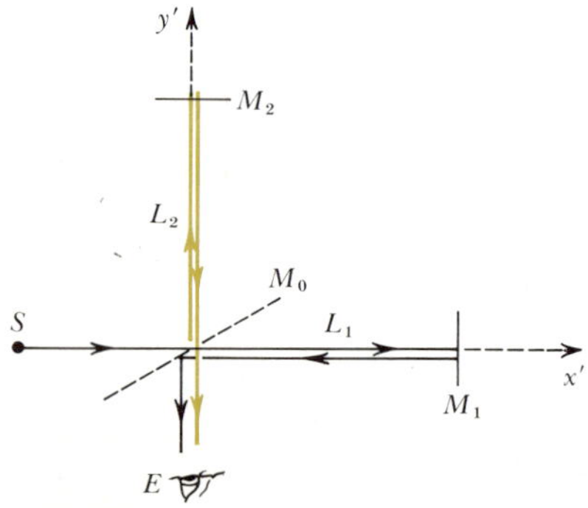

There is, however, one way to make indirect determinations of such a small time interval. The visible-light interference patterns discussed in Chapters 18 and 19 are due to the interference between two light signals originating at the same or identical sources but traveling through distances differing by approximately 1 wavelength. For visible light a path difference of 1 wavelength corresponds to a time difference of one period, or approximately 10^{-15} s. The American physicist A. A. Michelson (1852–1931) invented a sensitive optical instrument which not only provided interference fringes but, because the two paths were perpendicular to each other, was able to check on the drift of a laboratory through the conjectured ether. Michelson's *interferometer*, as it was called, is shown in Fig. 21-15. A light signal originating at source S is partly reflected by a half-silvered mirror M_0 to mirror M_2, a distance L_2 out the y' axis, and partly transmitted through M_0 to mirror M_1, a distance L_1 out the x' axis. The mirrors M_1 and M_2 reflect these signals back to mirror M_0 and thence to the observer's eye E on the negative y' axis. Michelson and E. W. Morley (1838–1923) used this instrument in 1881, and a more

FIGURE 21-15 Michelson's interferometer. Light from source S incident upon half-silvered mirror M_0 follows two paths to eye E: (1) light initially transmitted through M_0 is reflected back from mirror M_1 and then reflected by M_0 to eye; (2) light initially reflected by M_0 is reflected back by mirror M_2 and then transmitted through M_0 to the eye. These two paths, of lengths L_1 and L_2 and perpendicular to each other, are shown in a reference frame at rest with respect to the interferometer.

refined version in 1887, to check on the constancy of the speed of light. If the motion of observers affects the measured speed of light in the way common to other kinds of wave motion, then Michelson and Morley should have been able to detect this effect and to confirm the existence of the ether.

First consider the time Δt_1 it takes a light signal to move from M_0 to M_1 and back to M_0 when the laboratory is moving in the direction of the x' axis relative to the ether. In going from M_0 to M_1 the light signal moves in the same direction the laboratory moves, and one expects the speed of the signal relative to the laboratory to be $c - V$. But as this signal returns on M_0, the light and laboratory are moving in opposite directions and the speed is $c + V$. Thus, using nonrelativistic classical ideas,

$$\Delta t_1 = \frac{L_1}{c - V} + \frac{L_1}{c + V} = \frac{L_1}{c}\left(\frac{1}{1 - V/c} + \frac{1}{1 + V/c}\right)$$

$$= \frac{L_1}{c}\left(1 + \frac{V}{c} + \cdots + 1 - \frac{V}{c} + \cdots\right) \approx \frac{2L_1}{c}$$

In other words, to first order in the speed ratio V/c the out and back corrections cancel each other: there is no first-order change in Δt_1 relative to the time $2L_1/c$ which would be measured by observers at rest with respect to the ether. We must recalculate the effect using second-order terms

$$\Delta t_1 = \frac{L_1}{c}\left(\frac{1}{1 - V/c} + \frac{1}{1 + V/c}\right) = \frac{L_1}{c}\frac{1 + V/c + 1 - V/c}{1 - (V/c)^2}$$

$$= \frac{2L_1}{c}\left(1 + \frac{V^2}{c^2} + \cdots\right)$$

Now we find that there is a correction $(2L_1/c)(V/c)^2$, much smaller than the value $2L_1/c$ which would be measured if the laboratory were at rest relative to the ether. In Michelson's interferometer this time Δt_1 must be compared with the time Δt_2 it takes the other light signal to move from the mirror M_0 out to mirror M_2 and back to M_0. Because the laboratory is moving with speed V relative to the ether, this light path in frame R at rest with respect to the ether, is shown in Fig. 21-16. The time Δt_2 is easily determined from this figure

$$c^2(\tfrac{1}{2}\Delta t_2)^2 = V^2(\tfrac{1}{2}\Delta t_2)^2 + L_2^2$$

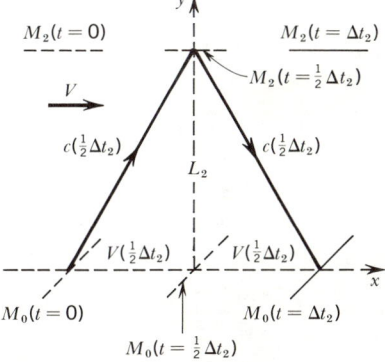

FIGURE 21-16 The Michelson interferometer of Fig. 21-15 moves with speed V with respect to the "ether" which is assumed to be the transmitting medium for the light signals. The path of the light reflected from M_0 to M_2 and back to M_1 is shown in this figure for observers at rest with respect to the ether when the interferometer moves parallel to the line from M_0 to M_1 in Fig. 22-15. The positions of the mirrors M_0 and M_2 are shown for three times: the beginning, midtime, and end of the interval Δt_2 during which the light signal is reflected from M_0 to M_2 and back to M_0.

$$\Delta t_2 = \frac{2L_2}{\sqrt{c^2 - V^2}} = \frac{2L_2}{c} \frac{1}{\sqrt{1 - (V/c)^2}}$$

$$= \frac{2L_2}{c}\left(1 + \frac{1}{2}\frac{V^2}{c^2} + \cdots\right)$$

The interference pattern seen by the observer's eye E is determined by the time difference

$$\Delta t_1 - \Delta t_2 = \frac{2L_1}{c}\left(1 + \frac{V^2}{c^2}\right) - \frac{2L_2}{c}\left(1 + \frac{1}{2}\frac{V^2}{c^2}\right)$$

Unfortunately it is impossible to be certain that L_1 is precisely equal to L_2; to determine this from the interference fringes would require knowing in advance that $V = 0$ or that the motion of the observer has no effect. Instead, one sets L_2 approximately equal to L_1 and observes the arbitrary position of the fringes. Then the entire apparatus is rotated 90°. L_1 and L_2 simply interchange their previous roles, and the new difference in time becomes

$$(\Delta t_1 - \Delta t_2)_{90°} = \frac{2L_1}{c}\left(1 + \frac{1}{2}\frac{V^2}{c^2}\right) - \frac{2L_2}{c}\left(1 + \frac{V^2}{c^2}\right)$$

As a result of this 90° rotation, there will be a shift in the position of any fringe relative to its position before the rotation. This shift will be determined by the time difference

$$(\Delta t_1 - \Delta t_2)_{0°} - (\Delta t_1 - \Delta t_2)_{90°} = \frac{2L_1}{c}\left(\frac{1}{2}\frac{V^2}{c^2}\right) + \frac{2L_2}{c}\left(\frac{1}{2}\frac{V^2}{c^2}\right)$$

$$= \frac{(L_1 + L_2)}{c}\frac{V^2}{c^2}$$

In their 1887 experiment, Michelson and Morley had path lengths equivalent to 11 m. The above shift corresponded to a time interval of $[(2.2 \times 10^1)/(3.0 \times 10^8)](10^{-4})^2 = 0.7 \times 10^{-15}$ s. The period of their light was about 2×10^{-15} s. Thus when the apparatus was rotated 90°, the fringes should have moved (or shifted) by as much as one-third the distance between adjacent fringes. Moreover their apparatus, which was mounted on a cement slab floating on mercury, was sufficiently sensitive to reveal shifts as small as one-hundredth the distance between adjacent fringes. Yet no motion of the fringes was detected. The experiment has been repeated many times under a variety of conditions (recent versions involve lasers as sources of light), and still no fringe shift has ever been observed. The experimental evidence is indeed convincing. Nevertheless, new and more precise experiments, one involving an Earth satellite carrying a hydrogen maser (atomic clock), are being developed. Such experiments should provide not only precise but very direct checks of this fundamental characteristic of nature, the universal constancy of the speed of light.

PROBLEMS

21-1 A rocket passes the Earth at a speed $0.6c$. According to observers on the rocket, city A is directly under the rocket at time $t = 0$, and city B is directly under the rocket 5.0×10^{-5} s later. According to observers on Earth, how long a time elapses between the arrival of the rocket at city A and its arrival at city B?

21-2 Observers on a train moving at $0.6c$ relative to a station platform see a man standing on the platform (1) pick up his suitcase and then (2) set it down. According to the observers on the train, the man held the suitcase for 4.0 s. By reading his own watch, how long does the man find that he held the suitcase?

21-3 A group of physics students are in a rocket moving at $0.9c$ relative to the Earth. A professor at rest on the Earth (*not* a physics professor) wishes to give the students an hour exam. When his watch reads 7:00 P.M., he signals the beginning of the exam, and when his watch reads 8:00 P.M., he signals the end of the exam. How long do the students, using their own watches, find that they have for the exam even after the finite travel time for the signals has been accounted for?

21-4 A cosmic-ray particle is observed to move straight toward the center of the Earth adjacent to a high-rise physics laboratory. The speed of the particle as observed by physicists at rest with respect to the laboratory is $0.6c$. (*a*) The particle is created (born) just as it passes a window ledge on the fifty-ninth floor and disintegrates (dies) just as it passes a window ledge directly below on the ninth floor. The physicists observe a time interval of exactly 1.0×10^{-6} s between these two events. What is the lifetime of the particle according to observers at rest with respect to the particle? (*b*) What is the vertical distance between adjacent window ledges, i.e., between floors, of the laboratory as measured by observers at rest with respect to the cosmic-ray particle?

21-5 A pion (lifetime of 1.5×10^{-8} s as measured in its own rest frame) was produced in a laboratory and observed to move 6.0 m in a straight line before decaying. (*a*) What was the speed of the pion as observed in the laboratory frame? (*b*) How long did the pion live according to the observers in the laboratory?

21-6 Astronauts pass an interplanetary space station with a speed 0.6c relative to the station. Then 12 s after the space ship passes the station as measured by observers at rest with respect to the station, a radio signal is sent from the station to the ship. (a) According to observers at rest with respect to the *ship*, how long a time interval elapses between the coincidence of the station and ship and the sending of the radio signal from the station? (b) How far from the station, as measured by observers at rest with respect to the *station*, is the ship when the radio signal is *sent* from the station? (c) How far from the ship, as measured by observers at rest with respect to the *ship*, is the station when the radio signal is *received* by the ship?

21-7 The distance from the Earth to Alpha Centauri is 4.3 light-years. A spacecraft makes this trip in 7.2 years as measured by observers on Earth. The trip started on the thirtieth birthday of one of the astronauts. (a) How old, according to the astronauts on the craft is this astronaut as he passes Alpha Centauri? (b) How far has the Earth traveled during this trip according to observers at rest with respect to the spacecraft?

21-8 A man expects to live an additional 50 years as measured in his own reference frame. If he wishes to reach a star 100 light-years away, as measured by observers at rest with respect to the Earth, just before dying, (a) at what speed must his space ship travel? (b) How long will it take him to get there according to observers on Earth?

21-9 The length of a thin straight rod is measured by observers moving relative to the rod with speed 0.5c in a direction parallel to the length of the rod. These observers using their own metersticks determine that the length of the rod is 30.0 m. What would the length of the rod be if measured by observers at rest with respect to the rod?

21-10 Astronauts in a fleet of spaceships passing the Earth at speed 0.8c relative to the Earth observe a one-lap race around the 300-m circumference of a circular track. The time for the winning athlete as measured by observers at rest with respect to the track is 50 s. (a) What do observers in the spaceships record as the winning time? (b) How long would the fleet of ships have to be, as measured by observers on Earth, in order for astronauts to be present for both the start and finish of the race? (c) Will the astronauts report that the Earth track is circular? (Assume the fleet of spaceships is wider than the track so that astronauts pass directly over and record the positions of all points of the track.)

21-11 Observers on a *station platform* see a train moving east with speed $V = 0.8c$. At time $t = 0$ they observe that the front end of the train coincides with the east end of the platform and the rear end coincides with the west end; they therefore conclude that the train has the same length as the platform, which they observe to be 70 m. (a) What is the length of the train as measured by observers at rest with respect to the *train*? (b) What is the length of the platform as measured by observers at rest with respect to the *train*? (c) Is the coincidence of the front end of the train and the east end of the platform simultaneous with the coincidence of the back end of the train and the west end of the platform according to observers at rest with respect to the *train*?

21-12 An evacuated tube connects the exit slit of a high-energy accelerator to the entrance slit of a particle detector. According to observers at rest with respect to the laboratory, the tube is 0.9 m long. A proton from the accelerator moves down the tube at a speed $0.6c$. (a) According to observers at rest with respect to the proton, what is the length of the tube? (b) According to observers at rest with respect to the laboratory, how long does it take the proton to move from one end of the tube to the other? (c) According to observers at rest with respect to the proton, how long does it take the tube to pass the proton?

21-13 A spacecraft moves from the Earth to a distant galaxy at speed $0.8c$ relative to Earth. The craft when at rest on the Earth was a cylinder 8 ft in diameter and 24 ft long. The longer dimension is always parallel to the motion. While the craft is moving, what are its diameter and length as measured (a) by observers on the craft and (b) by observers at rest with respect to the Earth?

21-14 A billiard ball moving east at 3.0 m/s approaches a second ball moving west along the same line at the same speed. What is the speed of the second ball as measured by observers at rest with respect to the first ball?

21-15 Repeat Problem 21-14 for two electrons each moving at $0.8c$ relative to the laboratory along the same line in opposite directions.

21-16 While a rocket ship is moving at $0.8c$ relative to the Earth, a particle is shot in the forward direction at $0.6c$ relative to the ship. What is the speed of the particle relative to the Earth?

21-17 Two spaceships recede from each other along a straight line. The speed of each ship as measured by observers on Earth is $0.9c$. What is the speed of the first ship as measured by observers at rest with respect to the second ship?

21-18 According to observers on Earth, a rocket moves east at a speed $0.5c$ directly toward a photon moving west at speed c along the same line. What is the speed of the photon according to observers on the rocket?

21-19 The electrons in the wire of Fig. 21-14 drift to the right with speed v. In each unit length of the wire there is positive charge of magnitude λ and negative charge of magnitude λ. Assuming that the magnitude of charge on an individual particle does not change when measured by observers in different reference frames, show that for a reference frame in which the electrons are at rest (a) the magnitude of the positive charge in each unit length of the wire is $\lambda/\sqrt{1-(v/c)^2}$, (b) the magnitude of the negative charge in each unit is $\lambda\sqrt{1-(v/c)^2}$, and (c) the net positive charge on each unit length is approximately $\lambda(v^2/c^2)$.

21-20 An east-west line of positive charges of uniform density λ C/m is moving at constant speed $v = 0.01$ m/s in the eastward direction. At one instant of time a particle of positive charge q at a radial distance R from the above line is also moving east at the same speed v. (a) Write an expression for the magnitude of the classical electric force $\mathbf{F}_e$ on q. Give your answer in terms of λ, q, and R (see Problem 13-17). (b) Write an expression for the magnitude of the classical magnetic force $\mathbf{F}_m$ on q. Give your answer in terms of λ, q, and R and v. (c) How can this classical magnetic force be accounted for in relativity theory?

RELATIVISTIC DYNAMICS

CHAPTER TWENTY-TWO

CHAPTER 22
RELATIVISTIC DYNAMICS

This chapter continues the revision of classical physics begun in Chapter 21. There expressions were developed which correctly describe space and time intervals for any relative speed of the moving observer. Our goal now is to find expressions of the same kind for the dynamical quantities momentum, force, and kinetic energy. As in Chapter 21, we can be confident, because of the success of our classical expressions, that at low speeds our new expressions will take on their familiar forms. Momentum, for instance, will reduce to $m_0 v$ and kinetic energy to $\frac{1}{2} m_0 v^2$. Here the subscript zero emphasizes $v \approx 0$. In classical Newtonian physics mass is assumed to be an intrinsic property of a particle, i.e., a property associated with the particle alone, irrespective of the presence of other particles or relative motion between the particle and the observer. That this assumption needs to be corrected will be our first result—mass, too, is relative.

22-1 MOMENTUM AND THE RELATIVITY OF MASS

By mass we have always meant that number by which you multiply the velocity of an object in order to get the vector quantity momentum that is conserved in all collisions. In Section 6-5, using the usual rule for connecting velocities in different reference frames, we found that conservation of the total momentum of the system is a *universal* law, i.e., one and the same statement of this law

[22-1] $\Sigma m_0 \mathbf{v} = \text{const}$

is valid in *all* inertial reference frames.

A second basic law that was found to hold universally is conservation of mass. It turned out from collision experiments that masses add arithmetically. Therefore the mass of a system of objects was just Σm_0, and the conservation law took the form

[22-2] $\Sigma m_0 = \text{const}$

This single formulation held in *all* inertial frames too; the *observer's* velocity nowhere entered into the definition of mass.

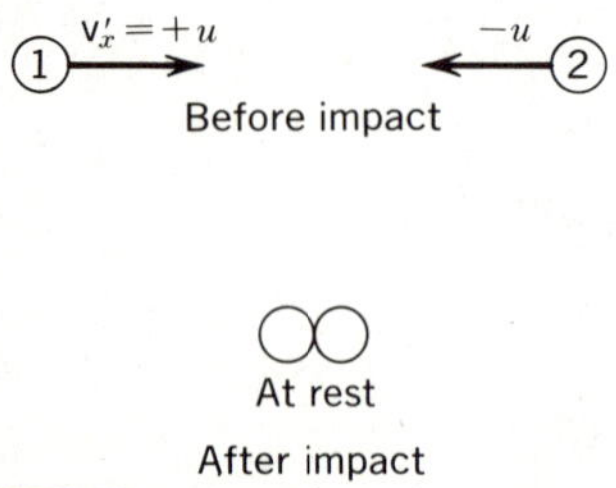

Now it does not take much to show, at least for objects moving with speeds close to that of light, that these two expressions for momentum and mass conservation are inconsistent with the correct velocity transformation [21-23] obtained in the last chapter. To see this, let's work out a specific example. Suppose a head-on, sticky collision between two identical objects takes place on a moving train. As shown in Fig. 22-1, the collision observed in the reference frame of the train R' is perfectly symmetrical, and therefore the composite

FIGURE 22-1 Head-on, sticky collision between two identical objects observed in the reference frame R' of a train.

body after the collision remains at rest. In this reference frame the speed of each object before collision is u.

Next, observe the collision in the reference frame R of the station platform. To keep the collision simple, we arrange it so that the velocity of the train past the platform is also $+u$. Then (see Fig. 22-2) for observers standing on the platform, the object on the right will be at rest initially, and the composite object after the collision will have the velocity $+u$. We can use the momentum- and mass-conservation principles [22-1] and [22-2] to determine the initial speed v of the left object

$$m_0(+v) + m_0(0) = M_0(+u)$$
$$m_0 + m_0 = M_0$$

From these two principles we find that $v = 2u$.

But we already know a correct way to calculate v in terms of u, irrespective of how fast the train is moving relative to the station. We use the relativistic velocity transformation [21-23], and for this special collision ($v'_x = u$ and $V = u$) we obtain

$$u = \frac{v - u}{1 - vu/c^2} \qquad [22\text{-}3]$$

Clearly if u and v are so small that vu/c^2 can be neglected compared to 1, we have the classical result $v = 2u$. However, if u is comparable to the speed of light, we get quite different results. For example, if $u = \frac{1}{2}c$, then $v = 0.8c$. (This is the same as Example 21-5.) This means that $v = 1.6u$, not $2.0u$. Something is very wrong! Evidently, the momentum- and mass-conservation laws are not universal; i.e., they do not have one and the same form for all reference frames—or else we have not formulated these conservation laws correctly.

A historic decision between these two alternatives was made by Einstein. He postulated the *principle of relativity: all laws of physics are universal.* Not just the speed of light but the expression of *every* physical law must be the same in all inertial frames. In other words, all inertial observers arrive at the same general statement of every law, each starting with his own measurements of space and time intervals.

Armed with this principle of relativity, one is forced to question the classical formulations of the momentum- and mass-conservation laws. Perhaps the assumption that mass is a constant property of an object, although successful in accounting for the motions of objects at speeds small compared to the speed of light, is not generally correct. Mass may be relative; it may depend on the speed of the

FIGURE 22-2 The collision of Fig. 22-1 as observed in the reference frame R of a station platform.

Classical momentum conservation and classical mass conservation are inconsistent with the correct relativistic velocity transformation.

The relativity principle

CHAPTER 22
RELATIVISTIC
DYNAMICS

Revised momentum and mass conservation assuming mass is relative

object relative to the observer. To check this possibility we continue to use m_0 as the symbol for mass when the object is at rest relative to the observer, but we shall use the more general symbol m when it is moving. The momentum- and mass-conservation laws, so modified, become:

[22-4]   Momentum conservation: $\Sigma m v = \text{const}$

[22-5]   Mass conservation: $\Sigma m = \text{const}$

Applied to our head-on, sticky collision between identical objects as observed in the platform reference frame (Fig. 22-2), our modified conservation laws yield

$$m(+v) + m_0(0) = M(+u)$$

and

$$m + m_0 = M$$

On eliminating the composite-body mass M from these two equations, we find that

[22-6] $$\frac{v}{u} = 1 + \frac{m_0}{m}$$

This modified result, of course, reduces to the classical result $v/u = 2$ if the observers on the platform measure the same mass m for the object moving with speed v as they measure for the mass m_0 at rest. On the other hand, this modified result allows for other values of v/u if m_0/m differs from 1. For example, we know from the correct relativistic velocity transformation that when $v = 0.8c$, $v/u = 1.6$. This means that the observers on the platform measure a *larger mass for the moving object* than for an identical object at rest. In our particular example, the object for which $v/c = 0.8$ has its mass increased by a factor $1/0.6$. From our experience with relativistic calculations, we might guess that the mass m for an object moving with speed v is related to the *rest mass* m_0 in general by

Relativity of mass deduced from the relativistic velocity transformation

[22-7] $$m = \frac{m_0}{\sqrt{1 - (v/c)^2}}$$

simply because we are familiar with the fact that the factor $\sqrt{1 - (v/c)^2}$ does equal 0.6 when v/c is equal to 0.8. To prove quite generally that [22-7] is indeed correct we solve the correct relativistic velocity transformation for this particular collision, [22-3], and obtain for v/u the expression

[22-8] $$\frac{v}{u} = 1 + \sqrt{1 - \left(\frac{v}{c}\right)^2}$$

The algebra involved is not particularly interesting, and we have relegated it to Example 22-3, but a comparison of [22-8] with the previous [22-6], which resulted from the application of our modified momentum- and mass-conservation laws to this collision, immediately establishes that [22-7] does give the correct variation of mass with the speed of the object relative to the observer.

The constant m_0 in [22-7] plays the same role as L_0 or Δt_0. By analogy, m_0 might be called the *proper mass* of a particle, but the name *rest mass* has become standard. In any reference frame which is in motion with respect to the particle the mass of the particle exceeds the rest mass.

Figure 22-3 shows m plotted against the particle speed v. At low speeds $m = m_0$, as required, but at speeds approaching c, m exceeds m_0 and approaches infinity for $v = c$. For example, at $v = \frac{1}{10}c$, an extremely high speed by all ordinary standards, m exceeds m_0 by a mere $\frac{1}{2}$ percent, but at $v = 0.98c$, m is 5 times greater than m_0.

Along with the new relativistic conception of mass comes a modified conception of momentum. The number by which we must multiply the velocity to get the conserved quantity, momentum, must be the relativistic mass, not the rest mass, so that

$$\mathbf{p} = m\mathbf{v} = \frac{m_0 \mathbf{v}}{\sqrt{1 - (v/c)^2}} \quad [22\text{-}9]$$

Of course, at low speeds this becomes identical to the classical definition $m_0 v$.

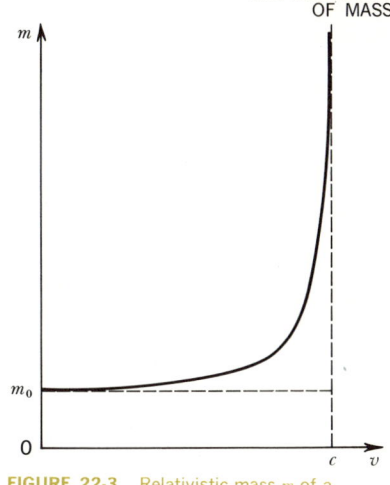

FIGURE 22-3 Relativistic mass m of a particle with speed v.

Relativistic momentum defined

EXAMPLE 22-1

A particle with rest mass m_0 first moves with the speed $0.000098c \approx 3.0 \times 10^4$ m/s, approximately the orbital speed of the Earth about the sun. Then the particle's speed is increased by a factor 10,000 so that it becomes $0.98c$. By what factor is the momentum of the particle increased?

SOLUTION

At the lower speed, $v/c = 10^{-4}$, so that the factor $\sqrt{1 - (v/c)^2}$ in [22-7] is essentially equal to 1 and $m \approx m_0$. For $v = 0.98c$, however,

$$\sqrt{1 - \left(\frac{v}{c}\right)^2} = \sqrt{1 - 0.98^2} = \sqrt{1 - 0.96} = \sqrt{0.04} = 0.2 = \frac{1}{5}$$

Therefore, at the higher speed, the momentum is increased not merely by the factor 10,000 in its speed but by the overall factor 50,000.

CHAPTER 22
RELATIVISTIC DYNAMICS

EXAMPLE 22-2 The individual particles in Figs. 22-1 and 22-2 have a rest mass m_0. (a) Using conservation of relativistic mass, express the rest mass M_0 of the composite particle in terms of m_0 and the speed u. (b) Using the same principle, find the relativistic mass M of the composite particle in terms of m_0 and the speed v. (c) M_0 and M of course describe one and the same composite particle; therefore, show that your answers to (a) and (b) are consistent with each other.

SOLUTION

a For the collision as seen by observers on the train (Fig. 22-1)

$$M_0 = m + m = 2m = \frac{2m_0}{\sqrt{1 - (u/c)^2}}$$

b For observers on the station platform (Fig. 22-2)

$$M = \frac{m_0}{\sqrt{1 - (v/c)^2}} + m_0$$

c We can express the result for M obtained in part (b) in terms of u rather than v by using the velocity transformation [22-3] in the rearranged form

[22-10]
$$\frac{v}{c} = \frac{2u/c}{1 + u^2/c^2}$$

to obtain after some algebra

$$\sqrt{1 - \left(\frac{v}{c}\right)^2} = \frac{1 - (u/c)^2}{1 + (u/c)^2}$$

Therefore the result in part (b) becomes

$$M = m_0 \left[\frac{1 + (u/c)^2}{1 - (u/c)^2} + 1 \right] = \frac{2m_0}{1 - (u/c)^2}$$

Then by substituting the result obtained for M_0 in part (a) into this last expression for M, we find

$$M = \frac{M_0}{\sqrt{1 - (u/c)^2}}$$

This is just the right relationship between M and M_0, the value of the mass measured by observers on the platform who see the composite object moving with speed u and the value of the mass measured by observers on the train, where the composite object is at rest.

EXAMPLE 22-3 Solve [22-3], the relativistic velocity transformation for Fig. 22-1 and Fig. 22-2 and obtain the expression for v/u given in [22-8].

SOLUTION First rearrange [22-3] into the form

$$u - \frac{vu^2}{c^2} = v - u$$

Multiplying both sides by c^2/v^3 and collecting terms, we get

$$\left(\frac{u}{v}\right)^2 - 2\frac{c^2}{v^2}\frac{u}{v} + \frac{c^2}{v^2} = 0 \qquad [22\text{-}11]$$

The reciprocal of the solution to this quadratic equation is

$$\frac{v}{u} = \frac{v^2/c^2}{1 \pm \sqrt{1 - (v/c)^2}}$$

Multiplying both the numerator and denominator of this last expression by $1 \mp \sqrt{1 - (v/c)^2}$, we finally obtain

$$\frac{v}{u} = 1 + \sqrt{1 - \left(\frac{v}{c}\right)^2}$$

where plus signs have been chosen to ensure that $v/u = 2$ when $v^2/c^2 \ll 1$. This result is [22-8].

22-2 RELATIVISTIC MOMENTUM

By definition the average net force $\langle F \rangle$ acting on a particle has the same magnitude and direction as the time rate of change of the particle's momentum

$$\langle F \rangle = \frac{\Delta(m\mathbf{v})}{\Delta t} \qquad [22\text{-}12]$$

This same definition applies for high-speed particles, where it is recognized that the mass is the relativistic mass. Thus, when Δt is small, we can write [22-12] as

$$\langle F \rangle = m\frac{\Delta \mathbf{v}}{\Delta t} + \mathbf{v}\frac{\Delta m}{\Delta t}$$

The second term on the right side has not appeared before since we regarded the mass of a particle to be a constant, that is, $\Delta m = 0$ for all Δt. But for high speeds the momentum change can arise both from a change in velocity and from a change in mass. Equivalently,

$$\langle F \rangle = m\mathbf{a} + \mathbf{v}\frac{\Delta m}{\Delta t} \qquad [22\text{-}13]$$

The first term on the right-hand side of [22-13] is the familiar mass times acceleration.

There is one situation in which a particle's mass does not change even though the particle is acted upon continuously by a force: a particle moving in a circle at constant speed. For example, when a particle with electric charge q moves at right angles to a magnetic field **B** at speed v, it traces out a circular path. The magnetic force has a constant magnitude $F_m = qvB$, and [22-13] yields

$$F_m = ma + v\frac{\Delta m}{\Delta t}$$

or

$$qvB = \frac{mv^2}{R} + 0$$

where R is the radius of the circle in which the particle travels with constant speed v, and hence with constant mass. We can simplify this relation to read

[22-14] $mv = qRB$

Note that the mass m in [22-14] is the *relativistic* mass. This relation can be used to test experimentally the mass variation with speed or, equivalently, the relativistic expression for momentum. All the quantities v, q, R, and B can be measured quite directly. It is found that m as computed from [22-14] is in complete agreement with the value of mass computed from [22-7]. Moreover, this agreement confirms our assumption that the electric charge of a given particle does not depend on the particle's speed.

22-3 RELATIVISTIC ENERGY

We know that the relativistic expression for a particle's momentum p is mv, where m is the relativistic mass. What is the relativistic expression for a particle's kinetic energy? Kinetic energy is defined as the work done by an external force in bringing the particle from rest to the speed v. Although we cannot tell in advance what form the relativistic expression for kinetic energy will take, we do know that for particle speeds much less than c it must reduce to the familiar $\frac{1}{2}m_0v^2$. It would be imprudent (and, as we shall see, wrong) to suppose that we can merely replace m_0 by m and then write the relativistic kinetic energy as $\frac{1}{2}mv^2$.

Let us square both sides of [22-7] and then multiply by $1 - v^2/c^2$. We then have

$$m^2 - \frac{m^2 v^2}{c^2} = m_0^2 \qquad [22\text{-}15]$$

Multiplying this equation by c^4 gives

$$(mc^2)^2 - (pc)^2 = (m_0 c^2)^2 \qquad [22\text{-}16]$$

where we have replaced mv by p.

Equation [22-16] contains squares of three terms—$m_0 c^2$, a universal constant for any given particle having rest mass m_0; pc, the relativistic momentum multiplied by c; and mc^2, which, like the other terms has the dimensions of energy (mass multiplied by speed squared). It seems likely that mc^2 is a physically important quantity because it combines with a momentum term to produce a universal constant. We can explore the meaning of mc^2 further by finding its form for particles with very low speeds. We can write

Relativistic energy defined

$$mc^2 = \frac{m_0 c^2}{\sqrt{1 - (v/c)^2}} = m_0 c^2 \left[1 - \frac{v^2}{c^2} \right]^{-1/2}$$

Expanding the last factor in this equation by the binomial expansion, we have

$$mc^2 = m_0 c^2 \left[1 + \frac{1}{2}\left(\frac{v}{c}\right)^2 + \frac{3}{8}\left(\frac{v}{c}\right)^4 + \cdots \right] \qquad [22\text{-}17]$$

If $v \ll c$, we can discard the term in $(v/c)^4$, so that

For $v \ll c$: $\qquad mc^2 \approx m_0 c^2 + \tfrac{1}{2} m_0 v^2$

or

$$mc^2 - m_0 c^2 \approx \tfrac{1}{2} m_0 v^2 \qquad [22\text{-}18]$$

The difference between mc^2 and $m_0 c^2$ is, at least for low speeds, the particle's kinetic energy! It is useful to designate the constant $m_0 c^2$ as the *rest energy* E_0 of the particle and mc^2 as the particle's *total energy* E, which depends on speed

$$E = mc^2 \quad \text{and} \quad E_0 = m_0 c^2 \qquad [22\text{-}19]$$

Then, if we denote the particle's kinetic energy by K and *assume* that [22-18] holds for all speeds, we have

$$E - E_0 = K \qquad [22\text{-}20]$$

In Example 22-4 we use calculus to show that the relativistic kinetic energy is indeed given by [22-20] for all speeds.

CHAPTER 22 RELATIVISTIC DYNAMICS

c EXAMPLE 22-4 A particle of rest mass m_0 initially at rest is acted upon by a force directed along the particle's displacement. The force does work on the particle and accelerates it to the final speed v. What is the final kinetic energy of the particle?

SOLUTION By definition, the kinetic energy K of a particle at speed v is equal to the work done by an external force in bringing the particle from rest to the speed v. Because our force and displacement are in the same direction,

$$K = \int F\, ds$$

Here we recognize that the force need not be constant, and we sum up, or integrate, the bits of work $F\, ds$ done over every small displacement ds. From [22-13]

$$K = \int \left(m\frac{dv}{dt} + v\frac{dm}{dt} \right) ds = \int (m\, dv + v\, dm)\frac{ds}{dt}$$

The derivative ds/dt represents the particle speed v, so that

[22-21] $\quad K = \int (mv\, dv + v^2\, dm)$

(Note now that if the mass were constant, i.e., if we did not take into account its relativistic change with speed, dm would be zero in [22-21]. Then the integral would reduce to $K = \int m_0 v\, dv = \tfrac{1}{2}m_0 v^2$, the Newtonian kinetic-energy expression.) The differentials of *two* variables appear in [22-21]; we must eliminate one of them. We can do so by first taking the differential of both sides of [22-15]

$$d(m^2) - d\left(\frac{m^2 v^2}{c^2}\right) = 0$$

$$2m\, dm - 2mv^2\frac{dm}{c^2} - 2vm^2\frac{dv}{c^2} = 0$$

or

$$(c^2 - v^2)\, dm = mv\, dv$$

Substituting for $mv\, dv$ in [22-21], we find

$$K = \int [(c^2 - v^2)\, dm + v^2\, dm] = \int_{m_0}^{m} c^2\, dm$$

Now the variable is the mass m, and the integral has the limits m_0 and m since the particle mass goes from m_0 to m under the action of the accelerating force. Finally,

$$K = mc^2 - m_0 c^2 = E - E_0$$

This is [22-20], rigorously established.

In Fig. 22-4 the total energy E is plotted against the particle speed v. Since $E = mc^2$ and $E_0 = m_0c^2$, this curve is of exactly the same form as that of Fig. 22-3 giving the relativistic mass m as a function of v. At very low speeds, for $v \ll c$ or equivalently for $K \ll E_0$, the kinetic energy can be written in the simply classical form $K = \frac{1}{2}m_0v^2$. At very high speeds, approaching that of light, the total energy E and the kinetic energy K go to infinity. This implies that it is impossible to bring a particle (at least one with a finite rest mass m_0) up to the speed c: to do so would require an infinite amount of energy. Thus, the speed c represents not only the speed of light signals in vacuum but also the limiting speed for particles.

Through the use of [22-19] we can rewrite [22-16] as

$$E^2 - (pc)^2 = E_0^2 \qquad [22\text{-}22]$$

For any given particle, the right side E_0^2 is a universal constant. Therefore, the left side of [22-22] must also be a universal constant for the same particle. As we go from one reference frame to another, E and p both change separately, but the quantity $E^2 - (pc)^2$ is invariant.

A new function of energy and momentum that has the same value in all reference frames

EXAMPLE 22-5

A particle moves with a speed v such that its kinetic energy is 4 times its rest energy. What is the speed v of the particle?

SOLUTION

The energy, rest energy, and kinetic energy are related (see [22-20]) by

$$E = E_0 + K$$

Thus for a particle with a kinetic energy 4 times its rest energy we have

$$K = 4E_0 \quad \text{or} \quad E = 5E_0$$

Using the definitions given in [22-19], this becomes

$$mc^2 = 5(m_0 c^2)$$

and the speed of the particle can be calculated using the relativistic expression for mass [22-9],

$$\frac{m_0 c^2}{\sqrt{1 - (v/c)^2}} = 5m_0 c^2$$

$$\left(\frac{v}{c}\right)^2 = 1 - \frac{1}{25} = \frac{24}{25} = 0.96$$

or

$$v \approx 0.98c$$

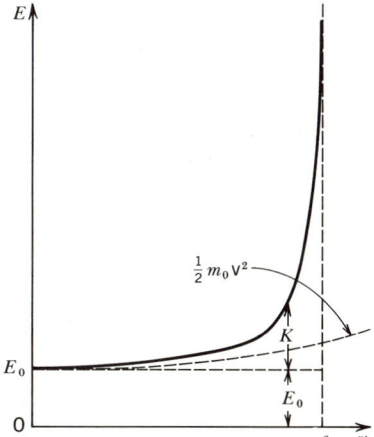

FIGURE 22-4 A particle's total energy $E = E_0 + K$, plotted as a function of its speed v. The classical kinetic energy $\frac{1}{2}m_0v^2$ departs from the relativistic kinetic energy at high speeds.

22-4 REST MASS AND ENERGY

We have found that when a particle is accelerated, both its energy and relativistic mass increase, and they are related by

[22-23] $$\Delta m = \frac{\Delta E}{c^2}$$

When the particle accelerates from rest, the change in energy ΔE is just the additional kinetic energy K. But, in addition, the particle always has an amount of energy E_0 associated with its rest mass.

This rather exotic notion of *rest energy* becomes more meaningful when we examine the mass changes in a simple collision between two particles, as illustrated in Figs. 22-1 and 22-2. Before impact, each particle has a mass m which exceeds its rest mass m_0 by virtue of its kinetic energy. After the perfectly inelastic collision the two-particle object is at rest. Using classical concepts, we would expect the mass of this composite object to be $2m_0$, but this is wrong. The correct rest mass M_0, as given by the relativistic mass-conservation principle, has the larger value $2m_0/\sqrt{1 - (u/c)^2}$ (see Example 22-2). A rest mass of $2m_0$ would be obtained only if both the colliding particles had been initially at rest. It is easy to understand why the two cases are different: the composite object with the larger rest mass has internal energy resulting from the initial kinetic energies of the particles in the inelastic collision, but the composite object formed from particles initially at rest does not have this additional energy. Evidently, then, rest energy accounts for all kinds of energy associated with an object at rest, including its internal thermal energy. Therefore, [22-23] also relates changes in *rest* mass to changes in *internal* energy. For example, a flask of gas weighs more when it is hot than when it is cold. Or, a spring is heavier stretched than relaxed because of its internal potential energy.

Changes in internal energy described by changes in rest mass

The concepts of rest mass and rest energy offer an interesting and useful new possibility, that of dispensing altogether with internal potential and kinetic energy. Instead, we can simply talk about an object's rest mass. Any change in internal energy will corrrespond to a change in this rest mass. Thus the process of heating a flask of gas or stretching a spring can be thought of as the transformation of one object into a new, more massive object. Such transformations, of course, require the absorption of energy. Other processes involve the emission of energy, whereby an object is transformed into a new object of smaller mass.

This may sound like a strange way to describe macroscopic bodies. If we were to cool the composite object formed as a result of the inelastic collision in Fig. 22-1 until it lost an amount of energy equal to the kinetic energy of the incident particles, we would not be likely to say that a new body had thereby been produced. But the resulting object would have an entirely different rest mass $2m_0$, so that we could quite properly call it a different entity.

In future chapters, when we describe atomic and subatomic objects of nature, we shall find ourselves using this type of description more and more frequently. There we shall meet with even more dramatic examples of the transformation between rest mass and kinetic energy, examples in which particles are either annihilated or created, the kinetic energy involved being accounted for by a loss or gain of the entire rest mass. Einstein considered the relationship between rest mass and energy to be the most significant result of his special theory of relativity.

EXAMPLE 22-6

Imagine that the incident objects in the inelastic collision of Fig. 22-1 are two air-track gliders each having a rest mass of exactly 1 kgm. Classically we would expect the mass of the composite object after the sticky collision to be exactly 2 kgm. How fast must the gliders move in order for the relativistic mass of this composite object to be 10^{-8} kgm (0.01 mgm) greater than the expected 2 kgm?

SOLUTION

In Example 22-2 we found the rest mass of the composite object to be

$$M_0 = \frac{2m_0}{\sqrt{1 - (u/c)^2}}$$

But for speeds small compared to the speed of light we can write this as (see [22-17])

$$M_0 \approx 2m_0\left(1 + \frac{1}{2}\frac{u^2}{c^2}\right)$$

Thus for incident particles of rest mass $m_0 = 1$ kgm and a composite particle of rest mass $M_0 = 2 + 10^{-8}$ kgm, we have

$$2 + 10^{-8} \approx 2 + \left(\frac{u}{c}\right)^2$$

or

$$u \approx 10^{-4}c = 3 \times 10^4 \text{ m/s}$$

For a barely detectable 0.01-mgm increase in rest mass, the incident gliders must move as fast as the Earth about the sun, or nearly 700,000 mph!

CHAPTER 22
RELATIVISTIC
DYNAMICS

EXAMPLE 22-7 In the collision described in Example 9-2 a spring is used as a bumper between the two colliding objects. The spring, which had an initial relaxed length of 0.50 m, was compressed to 0.4 m by the midtime of the collision, when the two objects came to rest. The initial total kinetic energy of 70 J was converted entirely to potential energy in the spring by this midtime. What was the increase in relativistic mass of the spring during this first half of the collision?

SOLUTION Using [22-23],

$$\Delta m = \frac{\Delta E}{c^2}$$

we get

$$\Delta m = \frac{+70 \text{ J}}{(3 \times 10^8 \text{ m/s})^2} = 8 \times 10^{-16} \text{ kgm}$$

This is about equal to the rest mass of 100 million (10^8) protons, but since an actual spring contains something like 10^{23} protons, the fractional change in mass is minute, indeed far below the level of detection. Later we shall consider submicroscopic systems where the *fractional* change in rest mass is much greater.

22-5 PARTICLES OF ZERO REST MASS

A zero-rest-mass particle moves with the single speed c but with any energy.

In Newtonian physics, rest mass and mass are the same thing. A "zero-rest-mass particle," with no momentum and no kinetic energy at any speed, would not be a particle at all but a mere spook. Relativity, however, opens up a new possibility.

Figure 22-5 shows the energy E plotted against v for several different values of the rest energy E_0. Included in this graph is the curve for a particle of zero rest energy and zero rest mass. For $m_0 = 0$, the energy is exactly zero for all values of v less than c. For $v = c$, however, we see that the energy line is vertical and that the energy E can assume *any* value. This means that a particle of zero rest mass is no longer an absurdity: the particle would exist, i.e., have mass and energy, only when in motion at the single speed c. Moreover, it could have any energy E for the speed c. Associated with this energy E, which is, of course, purely kinetic, would be a momentum p, as given by [22-22] with $E_0 = 0$

[22-24] $$p = \frac{E}{c} \quad \text{(for } m_0 = 0 \text{ and } v = c\text{)}$$

Thus, a particle with zero rest mass, one always in motion at the speed c, *can* have momentum and energy, and its momentum is just its energy divided by c.

This reminds us of the relation between the momentum and energy of an electromagnetic wave. In fact, [22-24] is precisely the same as [20-14] for an electromagnetic beam. When, on page 464, we tried classically to interpret the momentum p of the beam as mc and the kinetic energy E as $\frac{1}{2}mc^2$, we ran into a contradiction. But there is no contradiction from the relativistic point of view, according to which a momentum $p = E/c = mc^2/c = mc$ is inevitably associated with any form of energy E that travels at the speed c. Thus, the momentum of light, and along with it the pressure of light against surfaces, is actually predicted by relativity theory.

There is, however, a disquieting note. We have convinced ourselves that light or any other type of electromagnetic disturbance consists of *waves*; the phenomena of interference and diffraction could not be explained in any other simple way. But the relativistic relations from which we deduced $p = E/c$ apply to *particles*. The wave-vs.-particle controversy will be reopened in Chapter 23. Let us momentarily anticipate a (partial) victory for the particle camp and suppose that electromagnetic radiation does consist of zero-rest-mass particles called photons. We do this to be able to discuss a device of current interest, the photon rocket.

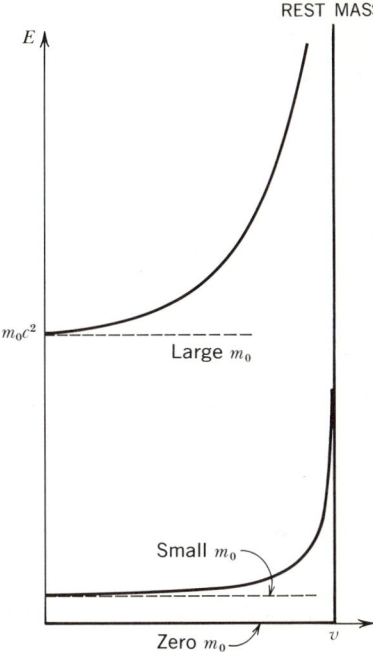

FIGURE 22-5 The total energy E as a function of speed v for three different values of the rest energy m_0c^2. Note that for $m_0 = 0$, the energy line is horizontal for v less than c and vertical at $v = c$.

EXAMPLE 22-8

The paradoxical relationship between the energy and momentum of an electromagnetic beam clarified

Since photons, traveling at the unique speed c, carry momentum, an object emitting photons in one direction, if isolated from external influence, will recoil in the opposite direction with an equal momentum magnitude, as required by the law of momentum conservation. One might, then, think of an object which sends out electromagnetic radiation in one direction as a particularly simple form of rocket.

The photon rocket

Despite the fact that not much momentum is carried off by electromagnetic waves of moderate energy, the photon rocket is actually more efficient than conventional rockets which eject particles from chemical reactions. To see this, consider the general relation between a relativistic particle's total energy E, its rest energy E_0, and its momentum p

$$p = \frac{\sqrt{E^2 - E_0^2}}{c}$$

For a given energy E, it is seen that p is biggest when E_0 is smallest, i.e., when $E_0 = 0$. This condition is met by photons, whose rest mass is zero. As compared to chemical fuels, then, light has literally more "bounce to the ounce."

Is the photon rocket technologically feasible? We can get an idea by examining a hypothetical voyage to a star 100 light-years away (as measured in a reference frame at rest with respect to the solar system and the distant star). If it is assumed that the rocket can give off radiation continuously and thus reach a speed comparable to c, the time-dilation effect will shorten the passengers' trip to much less than 100 years. For example, if $v/c = 0.98$, observers on earth see the voyagers take the dilated time $T = (100 \text{ light-years})/0.98c \approx 102$ years, but the voyagers themselves experience the proper time interval, which is (see [21-9])

$$T_0 = T\sqrt{1 - \left(\frac{v}{c}\right)^2} = (102 \text{ years})\left(\frac{1}{5}\right) \approx 20 \text{ years}$$

It is a straightforward matter to analyze how much mass must be radiated away from the rocket for it to attain the speed $v = 0.98c$. We begin with a rocket of rest mass m_0 and denote the *rest* mass remaining after it has achieved the speed v by fm_0. The "missing" mass is radiated. Therefore, energy conservation requires that

initial rocket E = final rocket E + radiated E

$$m_0 c^2 = \frac{fm_0 c^2}{\sqrt{1 - (v/c)^2}} + E_r$$

where E_r represents the total of radiated energy.

Momentum must also be conserved. If we assume, for simplicity, that the rocket takes off from rest in empty space, the system's total momentum, initially and thereafter, is zero. Therefore,

initial rocket p = final rocket p + radiated p

$$0 = \frac{(fm_0)v}{\sqrt{1 - (v/c)^2}} - \frac{E_r}{c}$$

where we recognize that the total momentum of radiated photons is $p_r = E_r/c$.

We eliminate E_r between these two relations to arrive at

$$m_0 c^2 = \frac{fm_0 c^2}{\sqrt{1 - (v/c)^2}} + \frac{fm_0 vc}{\sqrt{1 - (v/c)^2}}$$

Dividing through by $m_0 c^2$ and solving for f, we find when $v/c = 0.98$

$$f = \sqrt{\frac{1 - v/c}{1 + v/c}} = \sqrt{\frac{1 - 0.98}{1.98}} \approx \sqrt{\frac{0.02}{2}} \approx \frac{1}{10}$$

Thus, to shrink the travel time by a factor of 5, the rocket must radiate away 90 percent of its initial mass. This alone is a formidable obstacle: even such a strong radiator as the sun has lost less than 1 part in 10^4 of its mass over the last billion years. Suppose, though, for the sake of argument, that a means has been found to burn up mass efficiently in a photon rocket and that only one-tenth of the initial mass remains after the speed ratio $v/c = 0.98$ has been achieved. If the explorers are to come to rest on a planet of the distant star, the rocket must be slowed down to rest again. This means an additional reduction in rest mass by a factor of at least 10, and if there is to be a return trip to earth, the rest mass must be reduced by the factor 10 at least twice more, once on takeoff and again on landing. All told, then, a round trip using the most efficient means of rocket propulsion conceivable, attaining so high a speed that earth time is shrunk by the factor 5, requires a photon rocket whose final rest mass (the payload) is a mere 10^{-4} of the initial rest mass. Taking the residual rest mass of rocket and passengers to be merely 2,000 lb, we have that the rocket mass at the start must be 10,000 tons, or the weight of a hundred 747 superjets or of a battleship. Clearly, interstellar travel by photon rockets will not be realized tomorrow.

22-6 UNITS AND COMPUTATIONS IN RELATIVISTIC DYNAMICS

The only objects which can be raised to speeds closely approaching that of light are atomic particles, whose masses are tiny fractions of 1 kgm. It is useful, therefore, to have a special scale for measuring atomic masses in which the numbers come out simple.

By definition, the rest mass of a neutral atom of carbon 12 (the carbon isotope with 12 nuclear constituents) is exactly 12 mass units (abbreviated u). Since in 12 kgm of carbon 12 (1 kgm mole) there are 6.02×10^{26} atoms,

Mass unit of the atomic mass scale defined

$$1 \text{ u} = \frac{1}{12}\left(\frac{12 \text{ kgm}}{6.02 \times 10^{26}}\right) = 1.67 \times 10^{-27} \text{ kgm}$$

On the atomic mass scale all atomic particles have rest masses not very different from whole numbers, as shown in Table 22-1.

A convenient energy unit for atomic particles is the electron volt, the kinetic energy acquired by a particle of charge e as it falls freely

Particle	Rest mass, u
Electron	0.00055
Proton	1.008
Fluorine atom	18.998

TABLE 22-1

Mass and energy units related

through a potential difference of 1 volt. Thus, 1 eV = 1.60 × 10^{-19} J. Related units are the kiloelectron volt, 1 keV = 10^3 eV; million electron volt, 1 MeV = 10^6 eV; and the gigaelectron volt,[1] 1 GeV = 10^9 eV.

[1] The GeV is sometimes referred to as the BeV, or billion-electron volt, but this term is passing out of use because for most Europeans a billion means a million million (10^{12}) rather than a thousand million (10^9).

Let us find the rest energy corresponding to 1 u.

$$E_0 = m_0 c^2 = (1 \text{ u})c^2 = \frac{(1.67 \times 10^{-27} \text{ kgm})(3.0 \times 10^8 \text{ m/s})^2}{1.6 \times 10^{-19} \text{ J/eV}}$$

or

$$E_0 = 9.31 \times 10^8 \text{ eV} = 931 \text{ MeV}$$

Expressed in the units of MeV, the rest energies of the electron and proton are

Electron rest energy = 0.511 MeV $\approx \frac{1}{2}$ MeV

Proton rest energy = 938 MeV $\approx$ 1 GeV

Almost without exception, relativistic computations for atomic particles are simplest when masses are expressed in mass units and energies in electron volts (or related units). Furthermore, as shown in the following example, the most convenient unit for the momentum of a particle is MeV/c.

EXAMPLE 22-9 What is the momentum of an electron with a kinetic energy of (a) 10 eV and (b) 10 GeV?

SOLUTION *a* Since 10 eV is so very small compared to the rest energy of an electron, 0.511 Mev, we are safe in applying the simple classical relations

$$K = \frac{1}{2} m_0 v^2 = \frac{(m_0 v)^2}{2 m_0} = \frac{p^2}{2 m_0}$$

Then

$$p^2 = 2 m_0 K = \frac{2(m_0 c^2) K}{c^2} = \frac{2(0.511 \times 10^6)(10)}{c^2}$$

or

$$p = 3.2 \times 10^3 \text{ eV}/c$$

Notice how introducing the speed c and using the rest energy $m_0 c^2$ permitted the computation to be made easily. Of course, if we must have the momentum expressed in units of kilogram-meters per second, this is easy enough to get

$$p = 3.2 \times 10^3 \text{ eV}/c = \frac{3.2 \times 10^3 \text{ eV}}{3 \times 10^8 \text{ m/s}} (1.6 \times 10^{-19} \text{ J/eV})$$
$$= 1.6 \times 10^{-24} \text{ kgm-m/s}$$

b For an electron of 10 GeV kinetic energy the computation is still simpler, because the kinetic energy K is very large compared to $E_0 = 0.511$ MeV $= 5 \times 10^{-4}$ GeV. The general relation between E, p, and E_0 is given in [22-22]

$$E^2 = E_0^2 + (pc)^2$$

Since $E_0 \ll K$, $E = E_0 + K \approx K$. We discard E_0^2 in the above relation, obtaining

$$E \approx K \approx pc$$

or

$$p = \frac{K}{c} = 10 \text{ GeV}/c$$

We know, of course, that such an energetic electron has a speed very nearly equal to c.

EXAMPLE 22-10

An electron moves with a speed v such that its kinetic energy is 3 times its 0.5-MeV rest energy. What is its momentum in units of MeV/c?

SOLUTION

The energy of the particle is (from [22-20])

$$E = E_0 + K = E_0 + 3E_0 = 4E_0$$

Since this energy is related to the momentum and rest energy by [22-22]

$$E^2 - (pc)^2 = E_0^2$$

we have

$$16E_0^2 - (pc)^2 = E_0^2$$
$$pc = \sqrt{15}\, E_0$$

Thus

$$p = \sqrt{15}\, \frac{E_0}{c} = 3.9 \frac{0.5 \text{ MeV}}{c} = 1.9 \text{ MeV}/c$$

SUMMARY

If the rest mass of an object is m_0 when measured by observers at rest with respect to the object, then the (relativistic) mass m of the object when it moves with speed v relative to the observers is

[22-7] $$m = \frac{m_0}{\sqrt{1 - (v/c)^2}}$$

and the (relativistic) momentum of the object is

[22-9] $$p = mv$$

where m is the relativistic mass.

The (relativistic) kinetic energy of an object is

[22-20] $$K = E - E_0$$

where the total energy E and the rest energy E_0 are

[22-19] $$E = mc^2 \quad \text{and} \quad E_0 = m_0 c^2$$

The relativistic momentum and energy of an object are related by the expression

[22-22] $$E^2 - (pc)^2 = E_0^2$$

where E_0^2, the square of the rest energy of the object, is a universal constant; i.e., it is the same number for every observer, irrespective of his reference frame.

PROBLEMS

22-1 What is the mass of an electron moving at (a) 3 m/s, (b) 3×10^4 m/s, (c) $0.98c \approx 3 \times 10^8$ m/s?

22-2 What is the momentum (in units of GeV/c) of a proton moving at $0.8c$?

22-3 A train at rest has a length L_0. What is the observed length of this train when it is moving so fast that its mass is twice its rest mass?

22-4 The magnetic field in a cyclotron is 1.0 T. What are the radius and frequency of a proton's cyclotron orbit when its kinetic energy is (a) 10 MeV and (b) 600 MeV? (Use 900 MeV as the approximate rest energy of the proton.)

22-5 What must be the speed of a particle if it is to have a kinetic energy equal to its rest energy.

22-6 Draw three separate curves showing the variation of a particle's (a) mass, (b) momentum, and (c) energy with its speed.

22-7 For a proton moving at $0.8c$, what is (a) the total energy and (b) the kinetic energy?

22-8 A proton moves with a speed v such that its kinetic energy is 3 times its rest energy. What is the momentum of the particle? Give your answer in units of MeV/c.

22-9 What is the momentum of (a) an electron having a kinetic energy of 1 GeV, (b) a proton having a kinetic energy of 1 GeV, (c) a photon having an energy of 1 GeV?

22-10 At a \$250-million national laboratory in Illinois protons are accelerated to 500 GeV. (a) What is the mass of these accelerated protons relative to their rest mass? (b) What is their speed? (Express your answer in terms of Δ, where $v/c = 1 - \Delta$.) (c) What would be the energy of electrons moving at this same speed?

22-11 What are the approximate mass and kinetic energy of an electron (a) used in an x-ray machine and having a speed $0.94c$, (b) shot from a synchrotron with a speed $0.99999992c$, (c) shot from a linear accelerator at a speed $0.99999999985c$?[1]

[1] The costs of the machines are about \$50,000, \$200,000, and \$120 million, respectively.

22-12 A large laser can emit in 1 *millisecond* (ms) an intense pulse of light having an energy of 2,000 J! What would be the speed of a 1-gm object having the same momentum as this pulse of light?

22-13 The average *power* received at the top of the Earth's atmosphere from the sun is about 1 kW/m². (a) How much of the sun's mass reaches the Earth (radius, 6.4×10^6 m) each *year*? (b) Is the Earth getting more massive as a result of this influx of energy? *Hint:* Is the average temperature of the Earth changing?

22-14 Draw three separate curves showing approximately the variation of momentum p as a function of speed v for particle A having a rest mass of 2.0 kgm, particle B having a rest mass of 1.0 kgm, and particle C having zero rest mass. Label your three curves clearly. It is sufficient to determine each curve from as few as three points, estimating the other connecting points.

22-15 Consider a particle having energy E and momentum p as measured by observers in a reference frame R. Observers in another reference frame R' will measure an energy E' and a momentum p' where in general $E' \neq E$ and $p' \neq p$. Show that the combination $E^2 - (pc)^2$ is, nevertheless, the same for all observers, i.e., invariant.

22-16 (a) Derive a classical expression for the momentum of a particle in terms of its kinetic energy K and mass m_0. (b) Derive a relativistic expression for the momentum of a particle in terms of its kinetic energy K and rest energy E_0. (c) Show that your answer for (b) reduces to that of (a) if $K \ll E_0$.

22-17 Electrons are accelerated from rest by an electric potential difference of 1.0×10^6 V. Then the electrons enter a uniform magnetic field moving always at right angles to the magnetic field lines in a circle of radius 1.0 m. (a) What is the kinetic energy of such an electron? (b) What is the magnitude of its momentum? (c) What is the magnitude of the magnetic field?

22-18 Explain why our earlier expression ($p = E/c$) for the momentum of an electromagnetic disturbance having energy E is inconsistent with a classical particle model of light but consistent with a relativistic particle model.

PARTICLE PROPERTIES OF WAVES

CHAPTER TWENTY-THREE

The classical physics of Newton is the physics of everyday experience. Past the boundaries of that experience it begins to break down. We know already that the theory of relativity is needed to interpret observations of bodies moving at very high speeds. Now, in crossing the frontier of the very small, we shall find that the classical description must give way to *quantum theory,* which alone can account for many subatomic phenomena. Quantum theory involves such a radical shift in ideas that it deserves to be called more than a theory. It is really a whole new physics.

Strangely enough, one of the first quantum effects to be discovered, the photoelectric effect, was stumbled upon by Heinrich R. Hertz (1857–1894) in the course of his experiments with radio waves. These were the very experiments which, by confirming Maxwell's predictions, seeemed to establish once and for all that light was purely a wave phenomenon. But the photoelectric and other effects showed otherwise. Electromagnetic radiation actually displays properties which become comprehensible only if we imagine it to consist (once more!) of *particles*.

23-1 EXPERIMENTAL DESCRIPTION OF THE PHOTOELECTRIC EFFECT

The photoelectric effect is basically this: visible light, or electromagnetic radiation of higher frequencies, shines on a metal surface, with the result that electrons are emitted from the surface. This effect is used in all sorts of practical devices in which the interruption of a beam of light causes a change in an electric current, e.g., automatic door openers or the sound track of a motion picture.

The photoelectric effect: electrons released from metal surfaces by light

Hertz first observed the photoelectric effect in 1887 when he was studying the emission and absorption of electromagnetic radio waves. His apparatus was primitive: radio waves were generated by sparks jumping across a gap between a pair of conducting spheres, and the detector, placed some distance from the generator, consisted of a circular conducting loop also broken by a spark gap. Sparks in the detector arose from currents induced in the loop by the waves radiated from the generator. Hertz noticed that the spark jumped more readily across the detector gap when it was illuminated by the light from sparks in other parts of the laboratory. He found the enhanced sparking to be due to ultraviolet light (invisible light with wavelengths somewhat shorter than those of violet light), whether this radiation came from sparks or from any other source.

By 1889 it was established that charged particles were being released from the *negative* side of the gap. In 1897 J. J. Thomson (1856–1940) identified these particles as electrons by showing they had the same charge-to-mass ratio he had found in his cathode-ray experiments 2 years earlier.

Qualitatively, then, photoelectrons are emitted from certain metal surfaces by ultraviolet light. The energy brought in by the electromagnetic radiation appears, at least in part, as kinetic energy of the outgoing electrons. The quantitative aspects of the photoelectric effect relate to the intensity of the radiation (the energy transmitted per unit time through a transverse unit area), its frequency, and the chemical nature of the surface. Not surprisingly, it is found that the number of photoelectrons emitted per unit time is proportional to the intensity of the incident radiation. The more energy in, the more energy out. But very surprising is the observation first made in 1902 by P. Lenard (1862–1947). *The kinetic energies of the emitted electrons do not depend on the intensity of the incident light.* When the intensity is increased and consequently more energy reaches the surface per unit time, the number of emitted electrons increases but not the energy of a typical photoelectron. Indeed, whatever the intensity, no electron is found to have a kinetic energy exceeding a single fixed maximum, provided the incident radiation has one constant frequency. If, however, the frequency of the light does change, the maximum kinetic energy of the photoelectrons varies linearly with it. Furthermore, unless the incoming frequency exceeds a certain threshold value, no electrons are released from the surface, however great the intensity of the light.

Property 1: the more energy in, the more energy out

Property 2: photoelectron energy unrelated to incident-light intensity

Property 3: for a given material, the maximum photoelectron kinetic energy is determined only by the frequency of light.

Property 4: frequency threshold

This behavior is summarized in Fig. 23-1, where the maximum kinetic energy $(E_k)_{max}$ is shown as a function of the frequency ν of the incident monochromatic radiation for two different types of metals. Although the threshold frequency ν_0 is a characteristic constant for a given metal, it differs from one metal to another. For example, the threshold for potassium is about 5×10^{14} Hz (yellow light), but for copper it is about 10×10^{14} Hz (ultraviolet light). Despite the difference in threshold frequencies, the increase in $(E_k)_{max}$ for a given increase in ν is the *same for all metals*. Thus both straight lines in Fig. 23-1 have the same slope, and they can be represented by an equation of the form

$$(E_k)_{max} = -W + h\nu \qquad [23\text{-}1]$$

Here h, the slope, is a *universal* constant, with the measured value

$$h = 6.6262 \times 10^{-34} \text{ J-s} \qquad [23\text{-}2]$$

On the other hand, $-W$, the intercept on the E_k axis, varies from

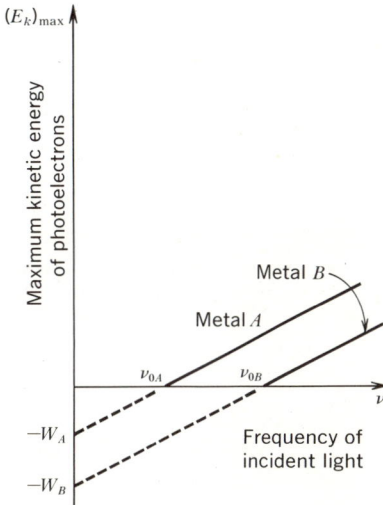

FIGURE 23-1 The maximum kinetic energy of photoelectrons as a function of the frequency of the monochromatic radiation illuminating the metal for two different metals. The threshold frequencies ν_{0A} and ν_{0B} differ, but both lines have the same slope.

metal to metal, just like the threshold frequency ν_0. Indeed, W can be related to ν_0. It is clear from Fig. 23-1 that when $(E_k)_{\max} = 0$, $\nu = \nu_0$. It therefore follows from [23-1] that

[23-3] $$W = h\nu_0$$

EXAMPLE 23-1 Photoelectrons are not emitted from the surface of a particular piece of metal if the wavelength of the incident light is greater than 5400 Å. (a) Calculate the value of W in [23-1] and express your answer in electron volts. (b) What is the maximum kinetic energy (in electron volts) of photoelectrons emitted from this same surface if the incident light has a wavelength of 2000 Å?

SOLUTION

a $$W = h\nu_0 = \frac{hc}{\lambda}$$

$$= \frac{(6.6 \times 10^{-34} \text{ J-s})(3 \times 10^8 \text{ m/s})}{(5400 \text{ Å})(10^{-10} \text{ m/Å})}$$

$$= 3.7 \times 10^{-19} \text{ J}\left(\frac{1}{1.6 \times 10^{-19} \text{ J/eV}}\right) = 2.3 \text{ eV}$$

b $(E_k)_{\max} = -W + h\nu = -W + hc/\lambda$

$$= -2.3 \text{ eV} + \frac{(6.6 \times 10^{-34})(3 \times 10^8)}{(2000 \times 10^{-10})(1.6 \times 10^{-19})}$$

$$= -2.3 + 6.2 \text{ eV} = 3.9 \text{ eV}$$

One more feature of photoemission must be mentioned. So far nothing has been said about *how long* it takes for the photoelectrons to appear, once the light has been turned on. For a very weak beam perhaps a sizable lag should be expected. Actual measurements reveal that *no matter how weak the incident light, some photoelectrons are released with no detectable delay*, i.e., in less that 10^{-9} s.

Property 5: photoelectrons are emitted without delay.

23-2 THE INADEQUACY OF CLASSICAL CONCEPTS

At first sight the photoelectric effect may appear more exotic, but fundamentally no more disturbing, than the phenomenon of thermionic emission. There, too, electrons are unbound from a metallic surface, the necessary energy this time coming from the reservoir of thermal energy of the hot metal. However, closer examination of the five properties listed in the last section shows that the concepts of classical physics prove inadequate and fundamentally defective when applied to the photoemission process.

Take the property that photoelectrons are emitted almost at once, even for very weak light. In the classical wave theory the energy of an electromagnetic beam is assumed to be spread continuously over the region of space occupied by the beam. Thus, when the beam is absorbed at a surface, its energy is transferred continuously; for an intense beam the energy extracted per unit time is large, and the rate at which energy leaves a beam of low intensity must be correspondingly small. Certainly no one electron in the metal surface can be dislodged until it has acquired enough energy to overcome its binding to the surface. This implies that for a very weak beam there ought to be a long delay before photoelectrons are emitted. As a matter of fact, this delay (see Example 23-2) should amount to days or years! Instantaneous emission is totally incomprehensible in classical terms.

Instantaneous photoelectric emission is incompatible with a classical interpretation.

EXAMPLE 23-2

As the temperature of a red-hot wire of copper is raised, the number of electrons boiled off from the surface in thermionic emission increases; indeed, by measuring the number of electrons released, one can deduce that each electron is bound to the surface with an energy of 7.2×10^{-19} J. Suppose that a very weak beam of light, with an intensity of only 10^{-10} W/m², shines on a copper surface (10^{-10} W/m² corresponds to about one ten-thousandth the intensity of starlight). How long would one expect to wait, on the basis of classical arguments, for the first photoelectrons to appear?

SOLUTION

Suppose, for simplicity, that the copper atoms are regularly spaced along the surface of the metal, like a checkerboard, the distance between adjacent atoms being 10^{-10} m. If the energy arriving per second at the surface is 10^{-10} J over an area of 1 by 1 m, then the energy arriving per second at one atom, with its area 10^{-10} by 10^{-10} m, is only 10^{-10} J $\times (10^{-10})^2 = 10^{-30}$ J. We assume that for each copper atom there is a single electron which can be released by photoemission. Then, if each electron is bound to the surface with an energy of 7.2×10^{-19} J, we should expect that enough energy is accumulated to release the electron only after a time of $(7.2 \times 10^{-19}$ J$)/(10^{-30}$ J/s$) = 7.2 \times 10^{11}$ s has elapsed. On the basis of this simple computation the waiting time is substantially more than 1,000 years, in violent contradiction to the observation that electrons, even for this very weak light intensity, are released instantaneously.

Next, look at the second property and its implications. Classical theory can provide no reason for the abrupt cutoff in the kinetic energies of the emitted electrons. Surely the speed of the fastest photoelectrons ought to increase with the intensity of the monochromatic illumination!

Classical theory cannot account for the threshold.

Perhaps most baffling from the classical point of view is property 4, the existence of a frequency threshold. It is not hard to conceive of a classical *energy* threshold for the release of electrons: this would just be the minimum energy needed by an electron to escape the attraction of nearby particles at the surface. But why can't one color of light furnish this minimum energy as well as another?

23-3 THE QUANTUM THEORY OF THE PHOTOELECTRIC EFFECT

In 1905 Einstein finally made sense out of the photoelectric effect by invoking a radical hypothesis, and curiously, 1905 was the same year Einstein introduced the theory of relativity, which so drastically altered classical conceptions of space and time. The relativity and quantum theories are the two great advances which underlie our understanding of all atomic phenomena. Both deal with matters far removed from common experience. As Einstein recognized, it would be dangerous—indeed, fallacious—to think that our commonsense notions necessarily had force in these remote domains.

Einstein and the quantum hypothesis of particlelike photons

According to classical electromagnetic theory, the energy of a light beam is *distributed continuously* throughout the beam. Moreover, a beam of light may have *any amount* of energy. Einstein saw that the photoelectric effect could be explained only if these propositions were discarded and replaced by the following:

1 The energy of electromagnetic radiation is *not* distributed continuously but *localized* in discrete *photons*.
2 The energy of electromagnetic radiation is *quantized*; i.e., electromagnetic energy can appear only in certain discrete amounts.

Specifically, in radiation of frequency ν each photon has precisely the energy

[23-4] $$E = h\nu$$

Consequently, the total energy of a beam of frequency ν and consisting of N photons is just $N(h\nu)$.

Photon energy $h\nu$; Planck's constant h

The constant h appearing in [23-4] is taken to be the universal constant appearing in [23-1]. This fundamental constant of the quantum theory is known as *Planck's constant*, since it was Max Planck (1858–1947) who first introduced the concept of energy quantization in 1900. In order to account for the thermal radiation from solid objects at high temperatures, Planck made the revolutionary assumption that the atomic oscillators in the radiating objects had quantized energies. Einstein's step was, if anything, even more drastic: he assumed the energy of electromagnetic waves to be quantized.

According to the quantum theory, the photoelectric effect is a straightforward example of energy conservation. The electromagnetic beam striking the photosurface is now supposed to consist of discrete, particlelike photons arriving one by one. As soon as the first photon reaches the metal surface, it interacts with one, and only one, electron and can give it as much energy as any electron will ever get from the beam. Leaving the beam on for a longer time or increasing its intensity will not change the amount of the energy transferred to a single electron, inasmuch as a change in the intensity of the light or in the time it illuminates the surface can change the number of the photons arriving at the surface but not the energy carried by each. For a very weak beam the electrons will be released less frequently on the average, but they will emerge essentially instantaneously and with the very same energy as for a highly intense beam of the same frequency. All this is exactly in accord with the observations.

Suppose that the relatively free electrons in the metal have a range of energies but that even the most energetic require an additional amount of energy W, called the *work function* of the metal, in order to escape from the metal surface against the attractive forces of the other charged particles. Obviously, then, no electron is released unless the energy it acquires in its encounter with an incoming photon is at least as large as this minimum energy W; that is, photoelectric emission is energetically possible only if the incoming photon energy $E = h\nu$ exceeds W. Thus, there is a threshold frequency ν_0 given by

$$W = h\nu_0$$

This is, of course, just [23-3].

It is now a simple matter to obtain [23-1]. Not all electrons in the metal are as energetically well suited for escape as the threshold ones; in general, the others will require an energy $\Phi > W$ before they can be released. If the incoming photon energy exactly equals Φ, all this incident energy is used up in dislodging the electron, so that the electron leaves the surface with zero kinetic energy; but if the photon energy exceeds Φ, there remains additional energy which is manifested as kinetic energy in the outgoing electron. In terms of energy conservation we have

$$h\nu \quad = \quad \Phi \quad + \quad E_k$$

| energy transferred from photon to photoelectron | energy lost by photoelectron in escaping from metal | kinetic energy of photoelectron after it leaves surface |

The electrons in the metal which are energetically best suited for emission, and hence must lose only the minimum energy W while escaping, will have the maximum kinetic energy after they leave the surface

$$h\nu = W + (E_k)_{\max}$$

This is, of course, just [23-1].

The photon hypothesis accounts for the photoelectric effect.

Thus, Einstein was able to derive the fundamental equation of the photoelectric effect from the quantum theory. The key idea was the quantization of the energy of the incident electromagnetic radiation. It is now clear why there must be a maximum kinetic energy of released photoelectrons and why the maximum kinetic energy changes in the same fashion for all metals for a given change in frequency. The particular type of metal determines how much of the photon's energy is used up in releasing photoelectrons, but the excess energy, whatever the material, goes entirely into the kinetic energy of the photoelectrons.

The quantum hypothesis accounts successfully for all the characteristics of the photoelectric effect which are incomprehensible in classical terms. That light is quantized and does consist of photons is established firmly, from the photoelectric and many other effects. Why the discreteness of light was not detected much earlier can be gathered from the following calculation. Let us find the number of photons hitting a postage stamp (area about 5 cm²) held about 6 ft (about 2 m) from a 100-W light bulb.

Supposing that 10 percent of the energy used by the bulb goes into emitted light, we have for the intensity I of the radiation at a distance of 2 m from the source

$$I = 0.1 \frac{100 \text{ W}}{4\pi(2 \text{ m})^2} \approx 0.2 \text{ J/(s)(m}^2) = 2 \times 10^{-5} \text{ J/(s)(cm}^2)$$

What is the energy of a single photon of this radiation? If we take the average frequency of light to be 10^{14} Hz, a photon of this frequency has an energy of

$$E = h\nu = (6.6 \times 10^{-34} \text{ J-s})(10^{14} \text{ Hz}) = 7 \times 10^{-20} \text{ J}$$

Thus, during each 1-s time interval the number of photons crossing an area of 1 cm² held at right angles to the rays of light is

$$\text{Number of photons} = \frac{2 \times 10^{-5} \text{ J/(s)(cm}^2)}{7 \times 10^{-20} \text{ J/photon}}$$
$$\approx 3 \times 10^{14} \text{ photons/(s)(cm}^2)$$

Thus roughly a thousand million million photons hit the stamp each second. Small wonder that the human eye does not distinguish individual arrivals. It can no more detect the individual photons from an ordinary light source than the hand held in front of a hair dryer can detect individual molecules of air. In both instances the essential granularity is masked by the enormous number of particles. (At the limit of sensitivity, however, the human eye *can* detect single photons from very feeble light sources; and, if the human ear were just *slightly* more sensitive, it would be able to detect individual molecular impacts.)

Photon energies are minute, and their number is enormous for an ordinary beam of light.

EXAMPLE 23-3

What is the energy in electron volts of a photon having a wavelength of 1000 Å (one lying in the ultraviolet region of the electromagnetic spectrum)?

SOLUTION

From [23-4] we have

$$E = h\nu = \frac{hc}{\lambda} = \frac{(6.63 \times 10^{-34} \text{ J-s})(3.00 \times 10^8 \text{ m/s})}{(10^{-7} \text{ m})(1.60 \times 10^{-19} \text{ J/eV})}$$

$$= 12.4 \text{ eV}$$

It follows at once that if the wavelength is 10 times greater, the photon energy is 10 times smaller. Indeed, one can write a general relation between the photon *energy in electron volts* and its *wavelength in angstroms* as

$$E = \frac{1.24 \times 10^4 \text{ eV-Å}}{\lambda}$$

EXAMPLE 23-4

A beam of monochromatic light of 6000 Å wavelength has an intensity of 3.0 kW/m²; that is, the electromagnetic energy passing through a unit transverse area per unit time is 3.0×10^3 J. (This corresponds to a 30-W red light source focused upon an area 10 by 10 cm.) What is the density of photons (the number per unit volume) in this beam?

SOLUTION

It is useful to derive first a general expression relating the photon density n to the intensity of the beam I and the energy $h\nu$ of each photon. In Fig. 23-2 photons travel parallel to the axis of a cylinder of length ct and cross-sectional area A. The time t then represents the time required for a photon moving at speed c to travel from one end of the cylinder to the other. If the number of photons per unit volume is n, the total energy per unit volume is $(h\nu)n$.

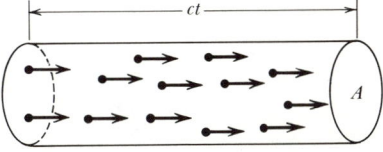

FIGURE 23-2

Therefore, within the cylindrical volume $A(ct)$ the total energy of photons is $(h\nu n)(Act)$. All the photons in the cylinder at any one instant will have passed through one end in the time t (and an equal number will have entered through the other end). The intensity, or photon energy *per unit time through the area A,* is then

$$I = \frac{h\nu n A ct}{At} = h\nu n c$$

or

$$n = \frac{I}{h\nu c} = \frac{I\lambda}{hc^2}$$

For the data given above we then have

$$n = \frac{(3.0 \times 10^3 \text{ J/s-m}^2)(6.0 \times 10^{-7} \text{ m})}{(6.6 = 10^{-34} \text{ J-s})(3.0 \times 10^8 \text{ m/s})^2}$$

$$= 3.0 \times 10^{13} \text{ photons/m}^3$$

This is equivalent to saying that there are about 30,000 photons of red light in a volume of *one cubic millimeter.*

23-4 PHOTONS AS PARTICLES

Even in the wave theory, electromagnetic radiation has some properties (energy and momentum) ordinarily associated with particles. As we found in Chapter 20, the electromagnetic momentum p and energy E are related by [20-14]

$$p = \frac{E}{c}$$

Recall that we arrived at a contradiction when we attempted (in Section 20-4) to interpret this relation in terms of *classical* mechanics: writing the momentum of the wave as $p = mv$ and the energy as $E = \frac{1}{2}mv^2$, we found that $v = 2c$, rather than $v = c$. But in terms of relativistic dynamics the relation $p = E/c$ finds a ready explanation: it connects the momentum p and energy E of a particle of zero rest mass, a particle which exists only when in motion at the single speed c. For zero rest mass we write momentum as $p = mv = mc$ and total energy as $E = mc^2$; clearly, there is no contradiction in $p = E/c$.

Quantum theory allows us to assert that such a particle actually exists and is exemplified by the photon. Thus, *a photon has properties of a particle whose rest mass is zero.* Like a molecule, which transfers energy by collision, a photon can relinquish its energy when interacting with electrons in the photoelectric effect. Although quantum theory attributes particlelike aspects to the photons of electromagnetic radiation, this does not mean that we can disregard the wave properties of light so convincingly demonstrated in such phenomena as interference and diffraction.

If electromagnetic energy is carried in discrete amounts by the individual photons, the same must be true of electromagnetic momentum. It, too, must be quantized. Combining the basic relations $E = h\nu$ and $p = E/c$, we obtain for the quantum of momentum

A photon has momentum h/λ.

$$p = \frac{h\nu}{c} = \frac{h}{\lambda} \qquad [23\text{-}5]$$

In 1923 Arthur H. Compton (1892–1962) demonstrated that a photon does indeed carry momentum, and of just the magnitude h/λ. We discuss his historic experiment in the next section.

Compton studied the interaction of high-energy x-rays with electrons in materials. When a loosely bound electron is struck by an x-ray photon, it acquires enough energy to escape from the atom. The maximum energy an electron can acquire is the total energy of the incident photon. For a typical x-ray photon, of wavelength 1 Å, the energy is

+ **23-5 THE COMPTON EFFECT**

The Compton effect: photons in collision with particles

$$E = h\nu = \frac{hc}{\lambda} = \frac{(6.6 \times 10^{-34} \text{ J-s})(3 \times 10^8 \text{ m/s})}{(10^{-10} \text{ m})(1.6 \times 10^{-19} \text{ J/eV})} = 10^4 \text{ eV}$$

Since this energy is about 1,000 times larger than the energy with which a typical electron is held to an atom, the loosely bound electron can be considered free, in the same sense that a baseball suspended by a weak thread can be considered free when struck by a heavy and rapidly swung bat.

Moreover, the energy of a 1-Å photon is still much smaller than the rest energy of an electron: 10 keV for the photon as against 511 keV for the electron rest energy. Thus, for an electron initially at rest it is proper to use nonrelativistic mechanics to analyze the interaction; an electron with a kinetic energy of only 10 keV moves with a speed much less than c.

What we have, in effect, is a collision of the billiard-ball type: a photon carrying energy $h\nu$ and momentum h/λ collides with an electron initially at rest. The electron acquires momentum mv and kinetic energy $\frac{1}{2}mv^2$ at the expense of the incident photon, and a

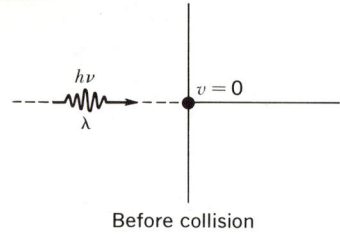

Before collision

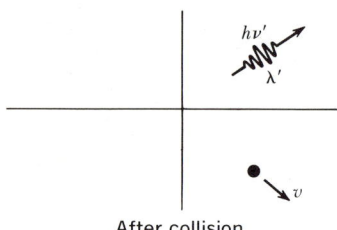

After collision

FIGURE 23-3 Photon colliding with an electron initially at rest.

A photon-electron collision analyzed in terms of energy and momentum conservation

scattered photon also leaves the collision site (see Fig. 23-3). The scattered photon differs in energy, momentum, and frequency from the incident photon. (The incident photon is annihilated in the collision, and the scattered photon is created.) Obviously, if the electron gains energy in the collision, the scattered photon must have less energy than the incident photon. Therefore, the scattered photon must have a frequency that is *less* than that of the incident photon.

According to classical electromagnetism, a wave is scattered as follows: the incident wave sets an electron in oscillation; then the absorbed energy is reradiated in other directions by the accelerated charged particle. Here the frequencies of the incident wave, the oscillating electron, and the scattered waves are all the same. Thus, the predictions of the classical and quantum theories are essentially different. Experiment shows that for x-rays the frequency of the scattered radiation actually is lower by a slight, but clearly detectable, amount. Indeed, Compton originally developed his theory of the quantized photon momentum to account for the observation of an otherwise unexplainable frequency shift.

We now compute the shift in wavelength by applying the principles of energy and momentum conservation to the collision between a photon and an electron. First, consider the simplest case in which the scattered photon moves in a direction opposite to that of the incident photon and the electron recoils in the forward direction (see Fig. 23-4). With the frequencies of the incident and scattered photons designated by ν and ν', respectively, energy conservation requires that

$$h\nu = h\nu' + \tfrac{1}{2}mv^2 \qquad [23\text{-}6]$$

energy before collision   energy after collision

$$\frac{hc}{\lambda} = \frac{hc}{\lambda'} + \tfrac{1}{2}mv^2$$

Similarly, momentum conservation gives

$$\frac{h}{\lambda} = mv - \frac{h}{\lambda'} \qquad [23\text{-}7]$$

momentum before collision   momentum after collision

The minus sign appears because the scattered photon travels backward.

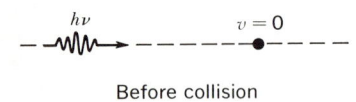

Before collision

After collision

FIGURE 23-4 Photon colliding head on with an electron; the electron recoils in the forward direction, and the scattered photon moves in the opposite direction.

We want to eliminate v from [23-6] and [23-7]. Equation [23-6] can be rewritten as

$$\frac{2hc}{m}\left(\frac{1}{\lambda} - \frac{1}{\lambda'}\right) = v^2$$

and [23-7] can be rewritten as

$$\frac{h}{m}\left(\frac{1}{\lambda} + \frac{1}{\lambda'}\right) = v$$

Combining the last two equations, we have

$$\frac{2hc}{m}\left(\frac{1}{\lambda} - \frac{1}{\lambda'}\right) = \left(\frac{h}{m}\right)^2\left(\frac{1}{\lambda} + \frac{1}{\lambda'}\right)^2$$

If the increase in wavelength, $\Delta\lambda = \lambda' - \lambda$, is small compared to either λ or λ', this last equation can be written as

$$\frac{2hc}{m}\frac{\lambda' - \lambda}{\lambda^2} = \left(\frac{h}{m}\right)^2\left(\frac{2}{\lambda}\right)^2$$

Here we have used the approximation $\lambda' \approx \lambda$ (except, of course, when taking the difference between these two nearly equal numbers). Finally, then,

$$\Delta\lambda = \lambda' - \lambda = 2\left(\frac{h}{mc}\right) \qquad [23\text{-}8]$$

Of course [23-8] applies only to photons scattered in one direction—backward. Next let us obtain the general result. We suppose that after the collision the electron moves in a direction making an angle θ with that of the incident photon while the scattered photon makes an angle ϕ (see Fig. 23-5). (The previous head-on collision corresponds to $\theta = 0°$ and $\phi = 180°$.) Energy conservation yields the same result as before, [23-6], but now there are separate equations for the x and y components of momentum. If we take the x direction to be that of the incident photon:

x direction: $\dfrac{h}{\lambda} = mv\cos\theta + \dfrac{h}{\lambda'}\cos\phi$ \qquad [23-9]

y direction: $0 = -mv\sin\theta + \dfrac{h}{\lambda'}\sin\phi$ \qquad [23-10]

The plan is first to eliminate θ from [23-9] and [23-10] and then, as before, to eliminate v. The final result will give $\Delta\lambda$ in terms of easily measured photon properties, λ, λ', and ϕ.

We write the identity

$$v^2 = (v\cos\theta)^2 + (v\sin\theta)^2$$

Solving for $v\cos\theta$ and $v\sin\theta$ in [23-9] and [23-10], we have

$$v\cos\theta = \frac{h}{m}\left(\frac{1}{\lambda} - \frac{\cos\phi}{\lambda'}\right)$$

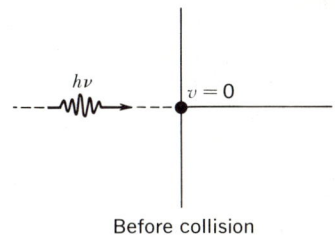

Before collision

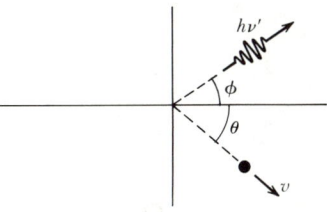

After collision

FIGURE 23-5 A Compton collision with the electron moving in a direction making an angle ϕ with that of the incident photon and the scattered photon making an angle ϕ.

A Compton collision analyzed for the general case

and

$$v \sin \theta = \frac{h}{m}\left(\frac{\sin \phi}{\lambda'}\right)$$

Substituting these expressions in the identity above then yields

$$v^2 = (v \cos \theta)^2 + (v \sin \theta)^2 = \left(\frac{h}{m}\right)^2\left[\left(\frac{1}{\lambda} - \frac{\cos \phi}{\lambda'}\right)^2 + \left(\frac{\sin \phi}{\lambda'}\right)^2\right]$$

After substituting this relation in [23-6], to eliminate v, we have

$$\frac{2hc}{m}\left(\frac{1}{\lambda} - \frac{1}{\lambda'}\right) = \left(\frac{h}{m}\right)^2\left[\left(\frac{1}{\lambda} - \frac{\cos \phi}{\lambda'}\right)^2 + \left(\frac{\sin \phi}{\lambda'}\right)^2\right]$$

Again we take $\Delta\lambda = \lambda' - \lambda$ as small compared to either λ or λ', so that $\lambda' \approx \lambda$. Then the above relation can be simplified to

$$\frac{2hc}{m}\frac{\Delta\lambda}{\lambda^2} = \left(\frac{h}{m}\right)^2\frac{1}{\lambda^2}(1 - 2\cos \phi + \cos^2 \phi + \sin^2 \phi)$$

or

[23-11] $$\Delta\lambda = \frac{h}{mc}(1 - \cos \phi)$$

As expected, our earlier special result, as given in [23-8], corresponds to the special case with $\phi = 180°$.

Equation [23-11], called the *Compton relation*, gives the wavelength shift $\Delta\lambda$ in terms of the scattering angle ϕ. The constant h/mc which appears in this relation is known as the *Compton wavelength*. For electrons it has the value

$$\frac{h}{mc} = \frac{6.6 \times 10^{-34} \text{ J-s}}{(9.1 \times 10^{-31} \text{ kgm})(3.0 \times 10^8 \text{ m/s})}$$
$$= 2.4 \times 10^{-10} \text{ m} = 0.024 \text{ Å}$$

Scattered photons have a longer wavelength.

Thus, if photons are incident upon loosely bound electrons, the photons scattered at right angles to the incident beam ($\phi = 90°$) have a wavelength that is greater by 0.024 Å, and photons scattered backward ($\phi = 180°$) have a wavelength shift of 0.048 Å. Note that these changes in wavelength do not depend on the wavelength of the incident radiation. For 1-Å x-ray photons, the wavelength shift is easily measurable; it amounts to 2.4 percent at 90° and 4.8 percent at 180°.

For scattering in the forward direction ($\phi = 0°$), $\Delta\lambda = 0$; this means that the incident photon does not collide with the electron at all, and the scattered photon is no different from the incident photon.

All these predictions have been confirmed in detail by experiments. The Compton shift occurs not only for x-rays but for all other wavelengths of electromagnetic radiation encountering free electrons. But for relatively long wavelengths, such as those possessed by photons of radio waves or even of visible light, the shift is so small as to be trivial. A 4000-Å photon of violet light scattered through 90° by a free electron becomes a photon whose wavelength is greater by 0.024 Å, but it is still basically a violet photon. The quantum and classical predictions agree, as they must when one deals with electromagnetic radiation of low frequency. Long-wavelength or low-frequency photons have a small energy and momentum, as compared to x-ray photons, and the quantization of energy and momentum, although just as real for such photons, is less conspicuous.

In summary, the photoelectric effect shows that the energy of an electromagnetic beam is quantized. The Compton effect confirms this and shows as well that the momentum of an electromagnetic beam is quantized. Since one must treat a photon of energy $h\nu$ and momentum h/λ as localized in space at the instant when it interacts with another object, in the fashion of a collision, there is no alternative to saying that photons, when interacting with other objects, have the properties of particles.

EXAMPLE 23-5

Photons of wavelength 0.012 Å are scattered by free electrons. What is the wavelength of those scattered photons whose angle of scattering is (a) 0°, (b) 60°, (c) 90°, (d) 120°, (e) 180°?

SOLUTION

Using [23-11],

$$\Delta\lambda = \frac{h}{mc}(1 - \cos\phi) = 0.024 \text{ Å } (1 - \cos\phi)$$

a   For $\phi = 0°$: $\Delta\lambda = 0$ $\lambda' = 0.012$ Å
b   For $\phi = 60°$: $\Delta\lambda = 0.012$ Å $\lambda' = 0.024$ Å
c   For $\phi = 90°$: $\Delta\lambda = 0.024$ Å $\lambda' = 0.036$ Å
d   For $\phi = 120°$: $\Delta\lambda = 0.036$ Å $\lambda' = 0.048$ Å
e   For $\phi = 180°$: $\Delta\lambda = 0.048$ Å $\lambda' = 0.060$ Å

EXAMPLE 23-6

A photon of wavelength λ moving directly east collides with a free electron initially at rest (see Fig. 23-6). The scattered photon of wavelength λ' moves directly west. Starting with *relativistic equations* describing conservation of energy and momentum and the relativistic expression relating the energy and momentum of a particle (see

[22-22]) show that the change in wavelength $\Delta\lambda = \lambda' - \lambda$ for the above photon is *precisely* that given by Eq. [23-8].

SOLUTION From relativistic energy conservation

$$h\nu + E_0 = h\nu' + E$$

or

$$\frac{hc}{\lambda} + E_0 = \frac{hc}{\lambda'} + E$$

or

$$E = E_0 + hc(\Delta\lambda/\lambda\lambda') \qquad [23\text{-}12]$$

From relativistic momentum conservation

$$\frac{h}{\lambda} = p - \frac{h}{\lambda'}$$

or

$$p = h\frac{\lambda + \lambda'}{\lambda\lambda'} \qquad [23\text{-}13]$$

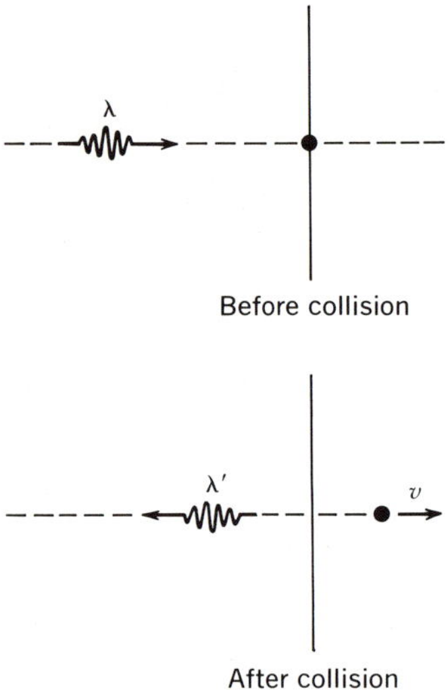

Here, we have three unknowns (E, p, and λ') and only two equations. But the energy and momentum of the free electron after the collision are, of course, related by [22-22]

$$E^2 - (pc)^2 = E_0^2$$

Substituting [23-12] and [23-13] into this relation, we get

$$\cancel{E_0^2} + 2E_0 \cancel{hc} \frac{\Delta\lambda}{\lambda\lambda'} + h^2 c^2 \frac{\Delta\lambda^2}{(\lambda\lambda')^2} - \frac{h^2 c^2}{(\lambda\lambda')^2}(\lambda + \lambda')^2 = \cancel{E_0^2}$$

or

$$2E_0\Delta\lambda + \frac{hc}{\lambda\lambda'}(\cancel{\lambda'^2} - 2\lambda'\lambda + \cancel{\lambda^2}) - \frac{hc}{\lambda\lambda'}(\cancel{\lambda^2} + 2\lambda\lambda' + \cancel{\lambda'^2}) = 0$$

or

$$\Delta\lambda = \frac{4hc}{2E_0} = \frac{2h}{m_0 c}$$

This exact, relativistic result, which did not involve the assumption $\lambda' \approx \lambda$, coincides with the approximate, nonrelativistic formula, [23-8].

EXAMPLE 23-7 When x-ray photons with a wavelength of 1.00 Å strike a carbon target, the scattered radiation is observed at 90° to the direction of the incident beam. As shown previously, the scattered radiation

FIGURE 23-6 Before collision / After collision

contains photons whose wavelength is 1.024 Å, but, in addition, there are also observed at this angle photons whose wavelength is just 1.00 Å. What is the origin of this radiation showing no shift in wavelength?

SOLUTION

The equation of the Compton effect relates the shift in wavelength $\Delta\lambda$ to the mass m of the particle responsible for scattering. *If the particle is an electron,* then, as shown above, $h/mc = 0.024$ Å. But if the mass is much greater than that of an electron, as would be the case for an entire atom, then the value of h/mc would be correspondingly smaller, as would also be $\Delta\lambda$ for any angle. In materials like carbon, only the outermost atomic electrons are so loosely bound as to be essentially free. A photon colliding with a tightly bound inner electron cannot move the electron without, at the same time, moving the entire atom. The photon must therefore be thought of as colliding with a particle whose mass is *thousands* of times greater than that of an electron. As a consequence, the scattered x-radiation from tightly bound electrons is effectively unchanged in wavelength. In general, then, there are *two* distinctive scattered wavelengths at each scattering angle: an unchanged wavelength arising from scattering from tightly bound electrons and an increased wavelength arising from scattering from essentially free electrons.

SUMMARY

In the photoelectric effect photons, of quantized energy $E = h\nu$, incident upon a metal surface release photoelectrons. The electrons, which lose the minimum amount of energy W in escaping from the metal, leave the surface with the maximum kinetic energy

$$h\nu = W + (E_k)_{\max} \qquad [23\text{-}1]$$

Here h is Planck's universal constant and W, the work function of the metal, is constant for a particular metal.

In the Compton effect photons, of localized energy $E = h\nu$ and localized momentum $p = E/c = h/\lambda$, are incident upon a loosely bound or free electron of mass m at rest. The result of the photon-electron interaction is a scattered photon of increased wavelength λ' and a deflected electron. The difference in photon wavelengths is

$$\Delta\lambda = \lambda' - \lambda = \frac{h}{mc}(1 - \cos\phi) \qquad [23\text{-}11]$$

where ϕ is the angle between the directions of the incident and scattered photons.

CHAPTER 23 PARTICLE PROPERTIES OF WAVES

PROBLEMS

23-1 Light of wavelength 5.0×10^{-7} m is incident upon a metal surface. The released photoelectrons, which have maximum kinetic energy, are just brought to rest by an electric potential difference of 1.0 V. What is the work function for the metal, i.e., the minimum energy required to release photoelectrons from the metal?

23-2 A lamp which provides intense illumination at ultraviolet frequencies is used to illuminate the surface of a flat piece of pure zinc metal. The emitted photoelectrons are observed. *Use Einstein's photoelectric theory to explain briefly your answers to the following questions.* (a) If the intensity (energy crossing a unit area per unit time) of the incident light beam is *increased,* what happens to the maximum kinetic energy of the emitted photoelectrons? (b) What happens to the number of photoelectrons emitted per unit time? (c) If a different lamp which provides illumination at *higher frequencies* but at the *same total intensity* as the first lamp is used, what happens to the maximum kinetic energy of the photoelectrons? (d) To the number of photoelectrons emitted per unit time? (e) If a thick *glass* lens is used to cover the entire front opening of the first lamp, what happens to the number of photoelectrons emitted per unit time? (Visible light but not ultraviolet light is transmitted through glass.) (f) To their maximum kinetic energy?

23-3 Radiation of frequency 3.2×10^{15} Hz is incident upon a metal surface. The work function of the metal is 3.3 eV. What is the kinetic energy (in electron volts) of the fastest electrons released from the metal?

23-4 What retarding potential in volts must be applied between the anode and cathode of a photoelectric tube in order to keep all the electrons released from the cathode from reaching the anode if the threshold frequency is (a) 10^{14} Hz and (b) 10^{15} Hz? (Again incident frequency is 3.2×10^{15} Hz.)

23-5 The photoelectric effect can be used to determine an experimental value for h. Light incident upon a metal surface releases photoelectrons. By applying a retarding electric field between the surface of the metal and a grid located a short distance above (and parallel to) the metal surface, it is possible to stop *all* photoelectrons from reaching the grid. The minimum electric potential difference between the surface of the metal and the parallel grid which stops all the photoelectrons is called the *stopping potential* V_0. What is the value of Planck's constant h if V_0 is 0.4 V for incident light of wavelength $\lambda = 5780$ Å and 1.4 V for $\lambda = 4050$ Å? (Use the known value of e.)

23-6 Show that it is impossible for an incident photon to transfer *all* its energy to the kinetic energy of a free electron initially at rest. *Hint:* Assume, tentatively, that the process *can* take place, and write equations expressing the conservation laws of momentum and energy in relativistic form. The process is shown to be impossible if both equations cannot simultaneously be satisfied.

23-7 Compute the photon energy in each instance: (*a*) radio waves from an AM station operating at a frequency of 1,500 kHz (*b*) radio waves from an FM station operating at a frequency of 100 MHz, (*c*) microwaves from a 10-cm radar speed trap, and (*d*) a 5000-Å lamp.

23-8 The retina of the human eye can detect visible light at intensities as low as 10^{-18} W. Approximately how many photons strike the retina each second under these conditions if the wavelength is 5000 Å?

23-9 At a distance of 4.0 m from a point monochromatic source of radiation the number of photons at any one time in a volume of 1 cm^3 is 10^8. (*a*) What is the number of photons per cubic centimeter 1.0 m away from the source? (*b*) What is the total number of photons radiated per second by the light source (assuming it to radiate uniformly in all directions)? (*c*) If all photons have a wavelength of 4000 Å, what is the power output of the light source?

23-10 (*a*) What are the momenta of a 100-eV electron (nonrelativistic) and (*b*) of a 100-MeV electron (relativistic)?

23-11 What is the wavelength of a photon scattered through 180° if (*a*) the incident photon of wavelength 0.002 Å is scattered by a free electron, (*b*) the incident photon of wavelength 0.004 Å is scattered by a free electron, (*c*) the incident photon of wavelength 0.002 Å is scattered by a free proton?

23-12 A photon moving east collides with a free electron initially at rest. The scattered photon moves directly north, and the electron moves in a direction making an angle θ with east. Show that, quite apart from the wavelength of the incident photon, angle θ can never exceed 45°.

ENERGY QUANTIZATION IN ATOMIC SYSTEMS: THE BOHR MODEL OF THE HYDROGEN ATOM

CHAPTER TWENTY-FOUR

CHAPTER 24
ENERGY QUANTIZATION IN ATOMIC SYSTEMS: BOHR MODEL

In Chapter 23 we found that electromagnetic radiation is quantized: in particular, the photoelectric and Compton effects can be explained only if electromagnetic waves of frequency ν have, not any value of energy, but only those values which are integral multiples of $h\nu$, the basic photon energy.

In turning to the atom and its structure, we again meet with energy quantization. Individual atoms have internal energies which are restricted to certain discrete values. Let us first look at the experimental evidence for this conclusion.

24-1 THE NUCLEAR ATOM

Not long after the atomic nature of matter had been established, it was recognized that the atoms themselves must have some sort of internal structure. It seemed clear from the work of J. J. Thomson and others that electrons were basic constituents of atoms, but since atoms are normally electrically neutral, there must also be constituents of positive charge. Indeed, after the discovery of natural radioactivity by Becquerel in 1897, the British physicist Ernest Rutherford (1871–1937) was able to show that this unusual radiation from matter fell into three categories: *gamma rays* (γ), which are high-energy, electromagnetic waves (even MeV photons rather than keV photons); *beta particles* (β), which are nothing other than energetic electrons; and *alpha particles* (α), which are particles with a positive charge of $2e$ and a mass 4 times that of a hydrogen atom, i.e., thousands of times more massive than an electron. Finding that helium gas was produced by substances emitting alpha particles, Rutherford concluded that these particles were simply helium atoms which lacked two electrons.

How is the positive charge of an atom distributed?

Granted that atoms contain positive and negative charges, the question arises how the charge is distributed. J. J. Thomson proposed that the atom was a sphere of continuously distributed positive charge, with the small electrons embedded throughout it (a "plum-pudding" atom). Rutherford, on the other hand, guessed that the positive charge might be concentrated in a relatively small volume near the center of the atom, called the *nucleus,* about which the electrons were somehow distributed (a "nuclear" atom).

Which picture of the atom is right—if, indeed, either one is? Certainly one cannot see an atom, but there remains the possibility of a scattering experiment, which consists in firing particles at an object to be examined and noting what happens to the particles, whether they are absorbed or deflected, and especially the degree to which they are absorbed or deflected. Actually, this is similar to looking at an ordinary object: the eye discerns what happens to electromagnetic particles, photons, which have been shot at the object.

Rutherford argued that if alpha particles were fired at a thin metal foil, the number of particles scattered at various angles would reveal whether the massive positively charged atomic constituents are distributed continuously throughout the atomic volume or restricted to a very small volume at the center. Any large deflections would be inconsistent with the plum-pudding model: neither the light electrons nor the diluted positive charge could significantly deflect heavy fast-moving alpha particles (see Fig. 24-1). Only an encounter with a concentrated charged body more massive than itself could send an alpha particle flying transversely or back toward its source (Fig. 24-2).

In 1913, at the suggestion of Rutherford, H. Geiger and E. Marsden performed an experiment using 7.68-MeV alpha particles as projectiles and gold foil 6.00×10^{-7} m thick as a target. They found that although most alpha particles passed straight through, some were deflected through large angles. Moreover, the relative number deflected through a given angle θ was properly accounted for by assuming that the small but strong scattering centers were positive point charges. The magnitude of each such positive charge was found to be equal to the magnitude of the electronic charge times the atomic number of gold (79). The mass of this positively charged object was essentially that of the entire gold atom, but its radius was at least a thousand times smaller than the radius of the atom. (Atomic radii are of the order of 10^{-10} m.) The evidence was overwhelming: alpha particles are deflected electrically by a charge localized in a volume having a radius no greater than 10^{-14} m. The scattering could be accounted for by the Rutherford nuclear model but not by the Thomson plum-pudding model.

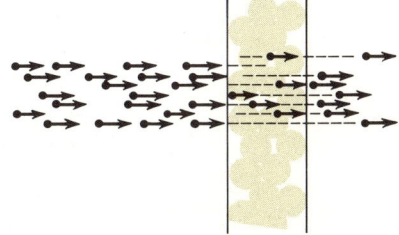

FIGURE 24-1 A beam of alpha particles fired at a thin metal foil. If the atom's positive charge is dilutely distributed throughout the entire atomic volume, as in the Thomson model, the alpha particles pass through the foil virtually undeflected.

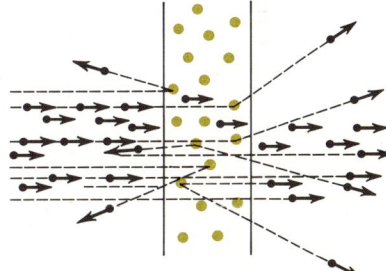

FIGURE 24-2 The scattering of alpha particles according to the Rutherford model. The alpha particles encounter massive positive point charges, and some are deflected through large angles.

The Rutherford nuclear model of the atom accounts for alpha-particle scattering.

24-2 THE INADEQUACY OF A CLASSICAL PLANETARY ATOMIC MODEL

If the positive charge of an atom is concentrated in a small central nucleus, it is not difficult to imagine how the electrons *might be* bound to the nucleus. They might orbit the much more massive nucleus as the planets orbit the sun, with the Coulomb electric force taking the place of the gravitational force. (Recall that the electric force between a proton and an electron is 10^{39} times greater than the gravitational force between these two particles.)

CHAPTER 24
ENERGY QUANTIZATION IN ATOMIC SYSTEMS: BOHR MODEL

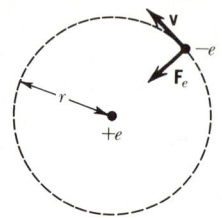

FIGURE 24-3 A classical model of hydrogen: the electron orbits the proton in a circle.

Attractive as this picture might be, it presents insuperable difficulties. It is enough to consider the simplest atom of all, hydrogen, the nucleus of which consists of a single proton and which has one electron.

If the electron is in planetary orbit, it continuously changes the direction of its velocity; i.e., it is continuously accelerated. According to classical electromagnetic theory, an accelerated charged particle radiates electromagnetic waves whose frequency is the same as that of the cycling particle. Thus, the hydrogen atom should radiate energy. Let us calculate, using classical arguments, the frequency of the radiation and check whether it corresponds to the measured frequency of light, e.g., the radiation from atoms in a neon sign or in a heated filament.

As a working approximation we may take the orbit to be circular, as shown in Fig. 24-3. The needed centripetal acceleration v^2/r is provided by the Coulomb attraction, $F_e = k_e e^2/r^2$. Hence, by Newton's law,

[24-1] $$\frac{k_e e^2}{r^2} = \frac{mv^2}{r}$$

in which m is the electronic mass. Since the nucleus of the hydrogen atom (the proton) is nearly 2,000 times more massive than the electron, we can properly take the nucleus to be at rest. This amounts to saying that we are viewing the hydrogen atom from a reference frame in which its center of mass is at rest.

Solving [24-1] for v, we find the orbital speed of the electron to be

[24-2] $$v = \sqrt{\frac{k_e e^2}{mr}}$$

This speed can be used, in turn, to determine the frequency f at which the electron orbits around the nucleus

[24-3] $$f = \frac{1}{T} = \frac{v}{2\pi r} = \frac{1}{2\pi}\sqrt{\frac{k_e e^2}{mr^3}}$$

The emitted electromagnetic radiation will have this same frequency; all that is needed now is a value for r.

A variety of independent measurements show that the hydrogen atom has a size, or diameter, of about 1 Å. Therefore, we take the orbital radius of the electron to be $\frac{1}{2}$ Å, or 0.5×10^{-10} m; and [24-3] yields $f = 7 \times 10^{15}$ Hz, which is the right order of magnitude for visible light.[1] So far, so good. But a slightly different value of r would

---

[1] The corresponding value of v is about 2.2×10^6 m/s, or less than $0.01c$. Our nonrelativistic calculation is therefore justified.

have given a slightly different frequency; and, classically, all values for the orbital radius are equally possible (provided they do not deviate too much from $\frac{1}{2}$ Å). Thus, the planetary model would lead us to believe that hydrogen atoms emit a *continuous spectrum* of light.

Classical planetary atomic model predicts a continuous emitted spectrum.

This is simply not so. Hydrogen radiates a set of *discrete* frequencies, a so-called *line spectrum* (see Fig. 24-4). In fact, all the elements have characteristic line spectra.

The first observations of discontinuous spectra were made in 1752 by the Scottish physicist Thomas Melville, using a prism, but the detailed observations came generally in the period from 1850 to 1900, after the development of very precise gratings capable of high resolution (see Section 18-9). Indeed, distinct groups, or *series*, of lines were found. The data, even for a single gas like hydrogen, became voluminous and remained uncorrelated. In 1885 a Swiss school teacher, Johann Balmer (1825–1898), hit upon an empirical formula which did give coherence to the data. For hydrogen, he found that all the emitted wavelengths could be summarized in the following quite simple equation:

$$\frac{1}{\lambda} = R\left(\frac{1}{n^2_{\text{smaller}}} - \frac{1}{n^2_{\text{larger}}}\right) \qquad [24\text{-}4]$$

where n represents an integral number and R is a constant equal to 1.1×10^7 m^{-1}, called the *Rydberg constant* (in honor of the Swedish spectroscopist J. R. Rydberg). In other words, for each emitted wavelength in the hydrogen spectrum, Balmer was able to find two integral numbers to which this wavelength could be related via [24-4]. Moreover, all the hydrogen lines in the family or series having *visible* light frequencies could be described by the common smaller integer, $n_{\text{smaller}} = 2$. The various lines of this series are then associated with the larger integers 3, 4, 5, ... This group of lines is now known as the *Balmer series*. In 1908, F. Paschen in Germany found several additional hydrogen lines at longer or infrared wavelengths (lower frequencies) which could be related to a smaller integer 3, together with the larger integers 4, 5, 6, ... And about the same time T. Lyman in the United States found a shorter-wavelength, or ultraviolet, hydrogen series which corresponded to the smaller integer 1 and larger integers 2, 3, 4, ...

FIGURE 24-4 Photograph of the line spectrum radiated by atomic hydrogen in the visible region of the electromagnetic spectrum. The lines are images of the slit used in illuminating the grating spectrometer (Section 18-9). Each separate image (line) corresponds to a discrete frequency (wavelength) of the incident light. (*Photograph from the Mount Wilson and Palomar Observatories.*)

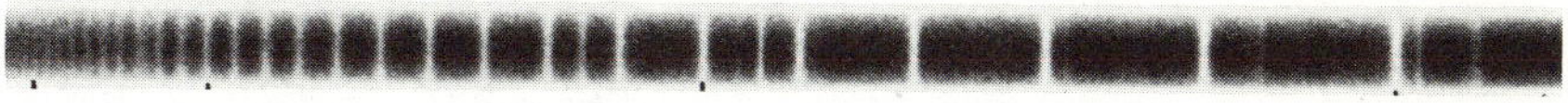

The empirical formula [24-4] described all these data with amazing precision. It was certain that the emission of light from atoms involves a beautiful regularity, a regularity which the planetary model can in nowise explain.

The classical atom collapses. There is a second, and graver, defect in the classical model. It cannot even account for the obvious fact that hydrogen atoms are stable, the same now as they were millions of years ago. The total energy of a radiating hydrogen atom must, by conservation of energy, be decreasing by precisely the amount of energy being radiated. Assuming, as before, a circular orbit around an immovable nucleus, we have from [24-2] for the electron's (and therefore, the atom's) kinetic energy

$$[24\text{-}5] \quad \mathsf{E}_k = \frac{1}{2}mv^2 = \frac{1}{2}\frac{k_e e^2}{r}$$

The potential energy of the atom can be obtained from [14-8]. With $q_2 = -e$ (the electron) and $q_1 = +e$ (the proton), we find

$$[24\text{-}6] \quad \mathsf{E}_p = \frac{-k_e e^2}{r}$$

The total energy of the atom is then

$$[24\text{-}7] \quad \mathsf{E} = \mathsf{E}_k + \mathsf{E}_p = -\frac{1}{2}\frac{k_e e^2}{r}$$

Note that the total energy E is negative. This means that energy must be added to, or work must be done on, the system to raise its energy to zero, i.e., to make the two particles infinitely separated ($\mathsf{E}_p = 0$) and both at rest ($\mathsf{E}_k = 0$).

From [24-7] it is seen that E decreases (becomes more negative) only if r decreases. That is, the planetary model requires that as energy is steadily radiated, the electron spiral inward toward the nucleus and at the same time the frequency of the radiation steadily rise (see [24-3]). But nothing of the sort occurs. Atoms do not collapse, and they radiate energy at fixed frequencies. The classical planetary model is just no good.

24-3 THE BOHR MODEL OF THE HYDROGEN ATOM

In 1911 a young Danish physicist, Niels Bohr (1885–1962), joined a group working with Rutherford at Manchester. Rutherford had only recently proposed his nuclear model of the hydrogen atom, but he was already aware that according to classical theory a planetary atom cannot be stable. Bohr had just come from the continent, where the new and controversial quantum theory of Planck and Einstein was causing much excitement. Bohr was the first to suggest that this theory, joined with the new nuclear model of the atom, might solve the perplexing problems of atomic stability and spectra. His contribution rested on a daring assumption: that classical electromagnetic theory, although correct for radiation from large-scale systems such as antennas, was not immediately applicable to small atomic systems.

Specifically, Bohr modified the classical theory by postulating that the electron of the hydrogen atom can actually remain in certain stable orbits *without* radiating (so-called *stationary orbits*) but that the atom *can* radiate a quantum of energy $h\nu$ when it makes a transition from one stationary orbit to another. He assumed the energy $h\nu$ of the emitted photon, by conservation of energy, to be just the difference in energy ΔE between the two stationary states of the atomic system

The Bohr theory and nonradiating stationary states

$$\Delta E = h\nu$$

For a transition from a higher- to a lower-energy state we can write

$$\Delta E = E_{higher} - E_{lower} = h\nu = h\frac{c}{\lambda} \qquad [24\text{-}8]$$

At once we see that the existence of nonradiative stationary states would explain, at least qualitatively, the discrete spectrum of hydrogen. For, if there is a restricted set of stationary states, the radiated frequencies ν and wavelengths λ in [24-8] cannot take on any values but only certain ones. What are these allowed values?

It is useful first to solve [24-8] for $1/\lambda$

$$\frac{1}{\lambda} = \frac{E_{higher}}{hc} - \frac{E_{lower}}{hc} \qquad [24\text{-}9]$$

Now let us compare this expression for the wavelengths of the emitted light with Balmer's empirical expression

$$\frac{1}{\lambda} = \frac{R}{n^2_{smaller}} - \frac{R}{n^2_{larger}}$$

The two equations are of the same form, and it is unlikely that this is mere accident. Indeed, remembering that atomic energies are negative, we can associate corresponding terms as follows:

$$E_{lower} = -\frac{Rhc}{n^2_{smaller}}$$

and

$$E_{\text{higher}} = -\frac{Rhc}{n^2_{\text{larger}}}$$

Both these expressions are summarized in the single formula

[24-10] $$E_n = -\frac{Rhc}{n^2}$$

In other words, the energies of the discrete stationary states of the hydrogen atom can be calculated by dividing the constant Rhc by the squares of integers. It is convenient to display the results in diagrams like that of Fig. 24-5. For $n = 1$, we have the lowest energy state, or level. The energy of this level is, of course, negative, since all the stable orbits correspond to negative total energies. The addi-

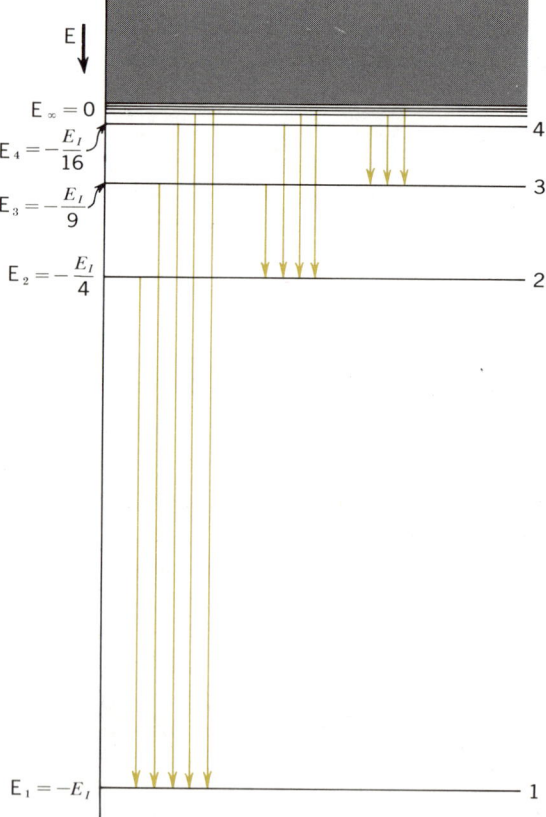

FIGURE 24-5 Energy-level diagram for hydrogen. Each level is identified by the quantum number n and by the atom's energy E.

tional stationary levels correspond to increasing values of n and higher (less negative) energies. If we think in terms of circular electron orbits, [24-7] shows that the lowest energy level corresponds to the orbit of smallest radius, while the highest level ($E = 0$) corresponds to an infinite radius (complete removal of the electron from the influence of the nucleus).

The energy required to separate an electron from the hydrogen nucleus is well known from chemical experiments; it is the ionization energy $E_I = 13.58$ eV. It should be possible to compute E_I from our general expression for the energy levels, and if the computed E_I turns out to be 13.58 eV, it would constitute a crucial check of Bohr's postulate. Normally a hydrogen atom is found in its lowest, or *ground, state*, $n = 1$. To ionize it, then, we must supply at least enough energy to raise the atom from the ground state to its highest level, $n = \infty$ (see Fig. 24-4). Thus the ionization energy has the value

$$E_I = E_\infty - E_1 = 0 + Rhc = Rhc \qquad [24\text{-}11]$$

Inserting the measured values of the Rydberg constant, Planck's constant, and the velocity of light, we obtain

$$E_I = \frac{(1.097 \times 10^7 \text{ m}^{-1})(6.625 \times 10^{-34} \text{ J-s})(2.998 \times 10^8 \text{ m/s})}{1.602 \times 10^{-19} \text{ J/eV}}$$
$$= 13.58 \text{ eV}$$

Hydrogen-atom ionization energy and radius computed from the Bohr theory

An agreement like this is nothing short of a triumph! Furthermore, the radius of the ground-state orbit, as computed from [24-7], comes out to be

$$r_1 = -\frac{k_e e^2}{2E_1} = 0.528 \text{ Å} \qquad [24\text{-}12]$$

again in excellent agreement with the known atomic radius of approximately 0.5 Å.

These calculations lent strong credibility to Bohr's marriage of the Rutherford nuclear model and quantum theory. The main achievement of the Bohr theory was its explanation of the discrete line spectrum. Of course, it was precisely for this purpose that the concept of stationary energy states was introduced in the first place.

The transitions responsible for the discrete spectral lines are represented by the vertical arrows superimposed on the energy-level diagram in Fig. 24-5. The various groups or series of lines are immediately apparent from this diagram. Thus, the transitions ending in the final state of energy E_2, corresponding to $n = 2$, produce the series

Energy-level diagram and transitions between stationary states

of spectral lines studied by Balmer, transitions ending in the stationary state having energy E_3 correspond to the Paschen series, and those ending in the E_1 state correspond to the Lyman series.

Bohr's stationary energy states can be used to explain absorption spectra as well as emission spectra. When white light having all possible frequencies passes through a gas, photons which have the characteristic frequency of a transition between two energy levels of the atomic system can be annihilated. Their energy is absorbed by an atom as it makes an upward transition, or quantum jump, to a higher or *excited energy state*. Photons having other frequencies are not absorbed. As a result the spectrum observed with a grating is continuous except for certain isolated black lines corresponding to the absence of light at the absorbed frequencies (see Fig. 24-6). Because the same energy levels are involved in absorption and emission, the frequencies of the black absorption lines on a colored background exactly match those of the colored lines on a black background seen in the emission spectrum.

EXAMPLE 24-1 A collection of hydrogen atoms, a gas of atomic hydrogen, is at room temperature, and all atoms are in their ground states. (*a*) What is the energy of the least energetic (longest-wavelength) photons that can be absorbed by the atoms of the gas? (*b*) What is the wavelength of these photons?

SOLUTION *a* We see from Fig. 24-5 that a quantum transition from the ground to the first excited state of hydrogen corresponds to the smallest energy difference ΔE. This transition requires that a photon of energy $h\nu$ be absorbed where, using [24-8], [24-10], and [24-11],

$$h\nu = E_2 - E_1 = \left(\frac{-E_I}{2^2}\right) - \left(\frac{-E_I}{1^2}\right)$$

$$= \frac{3}{4}E_I = \frac{3}{4}(13.58 \text{ eV}) = 10.2 \text{ eV}$$

Less energetic photons pass through the gas unabsorbed since

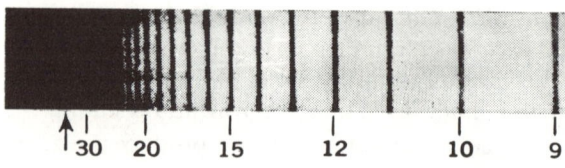

FIGURE 24-6 Photograph of the dark lines appearing in the absorption spectrum of sodium vapor. (*G. R. Harrison.*)

their energy does not match the energy difference between any quantized level of hydrogen and the lowest level. Of course, there *are* pairs of levels whose difference in energy is less than 10.2 eV, but here we have supposed that all atoms are initially in the ground state.

b The photon wavelength corresponding to 10.2 eV is computed simply from the general relation [24-8]

$$\Delta E = E_{\text{photon}} = h\nu = \frac{hc}{\lambda}$$

or

$$\lambda = \frac{(6.6 \times 10^{-34}\text{ J-s})(3 \times 10^8 \text{ m/s})}{(10.2 \text{ eV})(1.6 \times 10^{-19} \text{ J/eV})} \approx 1200 \text{ Å}$$

This is a photon lying in the ultraviolet region of the electromagnetic spectrum. Recall that the wavelength of the most energetic photons of visible light is about 4000 Å. Visible photons, therefore, have energies more than three times too small to be absorbed by hydrogen atoms in the ground state. Saying it differently, atomic hydrogen at room temperature is transparent to visible light. However, if the gas is heated or subjected to an electrical discharge, an appreciable fraction of the atoms will be found in the $n = 2$ or higher states. Consequently, photons can be absorbed which correspond to upward transitions from the $n = 2$ state. These transitions correspond to the lines of the Balmer series in the visible region.

EXAMPLE 24-2

Two hydrogen *atoms* in the ground state are traveling in opposite directions at the same speed. What is the minimum kinetic energy of either atom that will produce a completely inelastic head-on collision, one in which the two atoms are brought to rest?

SOLUTION

Up to this point we have supposed that an atom can make a quantum jump from one energy level to another only if a photon is emitted or absorbed in the process, with the photon energy exactly the same as the energy difference between the two atomic levels involved in the transition. The existence of such a photon ensures that energy is conserved. Quantum jumps, however, can also take place when the atom gains or loses energy in a collision with another particle. This effect was discovered by J. Franck and G. Hertz in 1914 by bombarding atoms of mercury with electrons. Since the bombarding particles lose energy only at certain discrete energies,

Quantum jumps in inelastic collisions

the Franck-Hertz experiment is an emphatic demonstration that atomic energies are quantized.

Here we are concerned with a couple of hydrogen atoms colliding head on. Then both atoms make a transition to the first excited state. This takes, all told, an energy of 2(10.2 eV), half of which is supplied by the kinetic energy of each atom. Thus, each atom must have a kinetic energy of exactly 10.2 eV before the collision. Note that in this inelastic collision the *total* energy of the system is conserved, even though its kinetic energy is not.

At approximately what temperature will such inelastic collisions of hydrogen atoms become frequent? The average translational kinetic energy per molecule of any gas at 300°K (room temperature) is $\frac{3}{2}kT \simeq \frac{1}{25}$ eV; therefore if the average kinetic energy of hydrogen atoms is to be about 10 eV, the gas temperature must be $(10 \text{ eV})(300°\text{K}/(\frac{1}{25} \text{ eV}) = 75{,}000°\text{K}$. At such an elevated temperature there will be many inelastic molecular collisions, whereby a substantial fraction of the atoms will be raised to excited states and then make downward transitions while emitting photons. In other words, hydrogen (or any other type of gas) will glow at high temperatures. That is why the sun shines. At low temperatures the colliding molecules have too little kinetic energy to cause transitions between stationary atomic or molecular states. The collisions cannot be even the least bit inelastic; they are perfectly elastic. The appropriateness of the basic assumption of perfectly elastic collisions in the kinetic theory of gases can be comprehended only by means of the quantum theory.

24-4 THE CORRESPONDENCE PRINCIPLE

To the extent that we have developed it in the previous section, Bohr's quantum theory is still partly empirical. The expressions it gives for the ionization energy E_I and the atomic radius r_1 of hydrogen are correct numerically, but those expressions involve the Rydberg constant R, which is not a fundamental constant of nature but merely a number which fits the spectroscopic data. In fact, Bohr's theory goes much further than this. It obtains E_I, r_1, and R *directly from the fundamental constants* h, c, k_e, and the electronic charge and mass. Essential to this achievement is a postulate or principle which belongs as much to philosophy as it does to physics. Let us briefly discuss Bohr's *correspondence principle*.

In the strictly classical planetary model of the hydrogen atom one takes the frequency of the emitted radiation to be simply the frequency f at which the electron orbits the nucleus. On the other hand, in the quantum theory the frequency ν of the emitted photon is computed quite differently, through the relation $h\nu = \Delta\mathsf{E}$. Based on the total evidence, the quantum theory is right, and the strictly classical theory is wrong. More properly, we should say that the classical electromagnetic theory is invalid for submicroscopic systems such as the hydrogen atom, but there is no doubt that it is quite satisfactory when applied to large-scale electromagnetic phenomena, e.g., ordinary electric circuits. This implies that the quantum theory is a refinement of classical electromagnetism. We then feel sure that the quantum theory, when applied to situations in which classical electromagnetic theory is adequate, *must* yield the *same* results as the classical theory does. In other words, when the more general theory is applied to situations in which the less general theory is known to work, the two theories should yield the same predictions. This very general assertion constitutes Bohr's correspondence principle. In the present instance it implies that the quantum frequency ν and the classical frequency f must be identical when the hydrogen atom is imagined to be so large in size that we can be sure that classical electromagnetism properly describes its behavior.

+ SECTION 24-4
THE CORRESPONDENCE PRINCIPLE

Bohr's correspondence principle: quantum results must correspond with classical results when conditions are macroscopic.

Before turning to the detailed comparison of ν and f, however, let us cite another situation where the correspondence principle operates. We know that relativistic dynamics is the only proper theoretical framework for comprehending phenomena involving speeds close to c. At the same time we recognize that Newtonian mechanics is perfectly satisfactory for describing low-speed situations. What the correspondence principle implies here is this: the classical Newtonian results *must* be obtained when relativistic mechanics is applied to low-speed phenomena. For example, the momentum of a high-speed particle is given by $m_0 v / \sqrt{1 - (v/c)^2}$. For low speeds, in which $v \ll c$, or $v/c \approx 0$, the relativistic relation for momentum reduces to $m_0 v$, which is, of course, the ordinary classical formula. Indeed, all the characteristic relativistic effects are described by relations in which the speed ratio v/c appears. When this ratio becomes negligibly small, the relativistic relations turn into the familiar relations of classical physics.

To return to the hydrogen atom, we have already found in [24-3] that the frequency of an orbiting electron is given by the classical theory as

$$f = \frac{1}{2\pi} \sqrt{\frac{k_e e^2}{mr^3}}$$

On the other hand, Bohr's quantum theory predicts for a transition between two adjacent energy levels, say n and $n-1$, radiation of frequency (see [24-7])

[24-13] $$\nu = \frac{E_n - E_{n-1}}{h} = -\frac{k_e e^2}{2h}\left(\frac{1}{r_n} - \frac{1}{r_{n-1}}\right)$$

Using our earlier relations between E_n, r_n, and n, we can express r_n in terms of r_1 by writing

$$E_n = -\frac{Rhc}{n^2} = -\frac{1}{2}\frac{k_e e^2}{r_n}$$

and

$$E_1 = -Rhc = -\frac{1}{2}\frac{k_e e^2}{r_1}$$

and dividing these last two equations

[24-14] $$r_n = n^2 r_1$$

Thus [24-13] can be written

$$\nu = -\frac{k_e e^2}{2hr_1}\left[\frac{1}{n^2} - \frac{1}{(n-1)^2}\right]$$
$$= -\frac{k_e e^2}{2hr_1}\left[\frac{n^2 - 2n + 1 - n^2}{n^4 - 2n^3 + n^2}\right]$$

We now suppose that n, and along with it the orbital radius r_n, is so big that it is safe to apply classical electromagnetism to the "atom." For such a large n, $2n \gg 1$ and $n^4 \gg 2n^3 - n^2$; therefore the above relation becomes

[24-15] $$\nu \approx +\frac{k_e e^2}{2hr_1}\frac{2}{n^3}$$

Equating f (from [24-3]) to ν (in [24-15]), as required by the correspondence principle, we have

Classical and quantum frequencies equated for a large hydrogen atom

$$\frac{1}{2\pi}\sqrt{\frac{k_e e^2}{mr^3}} \approx \frac{k_e e^2}{hn^3 r_1}$$

But the classical radius r is, to a very good approximation, simply $r_n = n^2 r_1$, because the stationary orbits get closer together, at least relative to each other, as n and r become larger (see Example 24-4). Therefore, taking r to be the same as $n^2 r_1$, squaring both sides of the equation above, and solving for r_1, we find

$$r_1 = \frac{h^2}{4\pi^2 k_e m e^2} \qquad [24\text{-}16]$$

Here is the radius of the hydrogen atom expressed in terms of fundamental constants alone. The corresponding theoretical values of the ionization energy and the Rydberg constant are, using [24-7] and [24-11],

$$E_I = \frac{1}{2}\frac{k_e e^2}{r_1} = \frac{2\pi^2 k_e^2 m e^4}{h^2} \qquad [24\text{-}17]$$

and

$$R = \frac{E_I}{hc} = \frac{2\pi^2 k_e^2 m e^4}{h^3 c} \qquad [24\text{-}18]$$

When we substitute numerical values for h, k_e, m, e, and π, we get

$r_1 = 0.528$ Å
$E_I = 13.58$ eV
$R = 1.0974 \times 10^7$ m^{-1}

all in excellent agreement with the observed values. Indeed, when the theoretical value of R is corrected to allow for the small motion of the nucleus, it agrees with the experimental value, 1.0967776×10^7 m^{-1}, to better than five significant figures!

The Bohr model is extremely successful in accounting for the spectra from hydrogen atoms but not so successful in explaining the behavior of more complicated atoms. As we shall see in the next two chapters, the Bohr theory inspired a more general quantum theory, which preserves two essential features: (1) the existence of discrete, stationary energy states and (2) the radiation or absorption of a quantum of energy as a transition is made between two of these states.

EXAMPLE 24-3

What is the binding energy of a negative muon (a particle having the same charge as the electron but a mass approximately 207 times as great) which has been captured by a proton to form an atom? What is the radius of the smallest Bohr orbit for this kind of atom (mu-mesic atom)? (Take the proton mass to be much greater than the muon mass.)

SOLUTION

The binding energy is equal to the magnitude of the energy of the lowest energy state, given by [24-7]

$$E_1 = -\frac{1}{2}\frac{k_e e^2}{r_1}$$

but the radius r_1 of the orbit corresponding to this state is (from [24-16])

$$r_1 = \frac{h^2}{4\pi^2 k_e m e^2}$$

where m is the mass of the small particle orbiting about the very massive nucleus. Thus, compared to the normal hydrogen atom, the new atom will have a radius r_1 approximately 207 times smaller, that is, $0.528 \text{ Å}/207 = 2.55 \times 10^{-3}$ Å. The binding energy of the new atom will therefore be 207 times larger than that of the hydrogen atom, that is, $(207)(13.6) = 2.82 \times 10^3$ eV $= 2.82$ keV.

EXAMPLE 24-4 Show that for the Bohr model of the hydrogen atom, the *fractional change* in the radii of two adjacent orbits approaches zero as the quantum numbers n for the orbits become very large.

SOLUTION The fractional change δ is given by

$$\delta = \frac{r_{n+1} - r_n}{r_n}$$

$$= \frac{(n+1)^2 r_1 - n^2 r_1}{n^2 r_1}$$

$$= \frac{2n+1}{n^2} = \frac{2}{n} + \frac{1}{n^2}$$

which approaches zero as n becomes very large. Thus, even though the radii become very large with increasing n, the *relative* spacing becomes much closer.

+ 24-5 QUANTIZATION OF ANGULAR MOMENTUM

All the predictions of the Bohr theory—the formulas for the allowed energies of hydrogen, for the frequencies of radiated and absorbed photons, and for the atomic radius—involve only fundamental physical constants and the integers n. We might well suspect that these integers are more than mere labels for the discrete energy levels. To uncover a deeper significance, let us rewrite the equation of motion [24-1] for the nth circular orbit

$$\frac{k_e e^2}{r_n^2} = \frac{mv^2}{r_n}$$

A little algebra brings this into the form

$$k_e e^2 m r_n = m^2 r_n^2 v^2$$

The right-hand side is recognized as the square of the quantity mvr, which is nothing else than the angular momentum, relative to an axis passing through the nucleus, of the orbiting electron. Since the nucleus remains essentially at rest, mvr is effectively the total angular momentum L of the atom. Thus

$$k_e e^2 m r_n = L^2$$

The radius is given by (see [24-14] and [24-16])

$$r_n = n^2 r_1 = n^2 \frac{h^2}{4\pi^2 k_e m e^2}$$

Substituting this into the previous equation, we obtain

$$k_e e^2 m \frac{n^2 h^2}{4\pi^2 k_e e^2 m} = L^2$$

or

$$L = n \frac{h}{2\pi} \qquad [24\text{-}19]$$

According to this important result, the *angular momentum* of the hydrogen atom is *quantized;* it occurs only in integral multiples of $h/2\pi$. Thus, the integer n is a *quantum number;* it tells how many angular momentum quanta $h/2\pi$ go with each of the stationary states.

Angular momentum is quantized in multiples of $h/2\pi$.

It is now possible to specify the stationary states of hydrogen very directly: only those states are permitted in which the atom's angular momentum is an integral multiple of $h/2\pi$. Accordingly, the Bohr quantum theory can be summarized as follows: (1) The electromagnetic radiation absorbed or emitted by an atom is quantized; or, only integral numbers of photons, each of energy $h\nu$, are ever generated or absorbed by an atomic system. (2) The angular momentum of an atomic system is quantized; or, the system's angular momentum is always an integral multiple of $h/2\pi$. This formulation brings out the crucial difference from classical theory: quantities such as energy and angular momentum, which in classical physics were assumed to take on any value, are found, in fact, to be quantized. Quantization exists for large-scale as well as for submicroscopic systems, but in the former case the effects are almost always totally unobservable. Thus, the angular momentum L of Mars as it revolves about the sun is so very large compared to $h/2\pi$ that we never notice that the orbit is such as to make $L/(h/2\pi)$ a whole number. To put it differently, at the enormous quantum numbers of macroscopic systems, the allowed states are so finely spaced, relative to the magnitude of the whole effect, that they are indistinguishable from a

CHAPTER 24
ENERGY QUANTIZATION
IN ATOMIC SYSTEMS:
BOHR MODEL

continuum. It appears that the system can take on any value of energy or angular momentum.

For large systems, then, quantum effects are minute and classical physics is valid. How large is "large"? The answer is determined by the size of Planck's constant h, which characterizes the quantum of electromagnetic radiation $h\nu$ and the quantum of angular momentum $h/2\pi$. One might say, in the spirit of Bohr's correspondence principle, that quantum effects can be disregarded whenever h can be treated as negligibly small. Or, classical physics is the correspondence limit of quantum theory as h approaches zero.

Classical physics, the correspondence limit of quantum theory with h negligibly small

SUMMARY

In the Bohr model, the hydrogen atom exists in discrete stationary states of total energy E_n without radiation of energy. When the system makes a transition from one energy state to a lower (higher) energy state, energy is radiated (absorbed) in the amount $h\nu$

[24-8] $\quad h\nu = E_{\text{higher}} - E_{\text{lower}}$

The discrete wavelength of the radiation is given by

[24-4], [24-9] $\quad \dfrac{1}{\lambda} = \dfrac{\nu}{c} = R\left(\dfrac{1}{n^2_{\text{smaller}}} - \dfrac{1}{n^2_{\text{larger}}}\right) = \dfrac{E_{\text{higher}}}{hc} - \dfrac{E_{\text{lower}}}{hc}$

where R is the Rydberg constant.

The electron is assumed to move in circular orbits about the massive nucleus. The energies of the stationary states are given by

[24-7] $\quad E_n = -\dfrac{1}{2}\dfrac{k_e e^2}{r_n}$

where

[24-14] $\quad r_n = n^2 r_1$

and r_1, the radius of the orbit for the lowest energy state ($n = 1$), can be determined from the ionization energy E_I or the Rydberg constant

[24-11], [24-7] $\quad E_I = Rhc = E_1 = \dfrac{1}{2}\dfrac{k_e e^2}{r_1}$

or strictly from fundamental constants [24-16].

The angular momentum L of a hydrogen atom in a stationary state of energy E_n is an integral multiple of $h/2\pi$

$$L = n\frac{h}{2\pi} \qquad [24\text{-}19]$$

PROBLEMS

24-1 Alpha particles having a kinetic energy of 7.7 MeV were used in the experiments of Geiger and Marsden in bombarding atoms of gold, the nucleus of which has a positive charge 79 times greater in magnitude than that of electron. Consider the special collision in which an alpha particle of this energy strikes a gold nucleus head on and the alpha particle is momentarily brought to rest. (a) What is the electrostatic potential energy of the system consisting of alpha particle and gold nucleus at this instant? (Assume the gold nucleus to remain essentially at rest.) (b) What is the minimum distance separating the alpha particle and gold nucleus in such a collision? (c) Suppose that the beam of alpha particles is replaced by a beam of protons also with a kinetic energy of 7.7 MeV. What is now the system's electrostatic energy and (d) closest approach distance for a head-on collision?

24-2 Calculate the wavelengths and energies in electron volts for the longest-wavelength (lowest-frequency) lines of the hydrogen series with (a) $n_{smaller} = 1$ (Lyman), (b) $n_{smaller} = 2$ (Balmer), and (c) $n_{smaller} = 3$ (Paschen).

24-3 In making a transition from the ground state what is (a) the longest wavelength that can be absorbed by hydrogen atoms? (b) the shortest wavelength?

24-4 A hydrogen atom is in its ground state. According to the Bohr theory, what is the ratio of the electron's speed to that of light?

24-5 A continuous spectrum of ultraviolet and visible light is sent through a collection of hydrogen atoms in a gas at *room* temperature. Show that the absorption spectrum will consist only of dark lines in the Lyman series.

24-6 The longest wavelength of hydrogen light in the visible region of the electromagnetic spectrum is 6.563×10^{-7} m ($n_{smaller} = 2, n_{larger} = 3$). What is the wavelength of the adjacent line in this series?

24-7 A collection of hydrogen atoms all initially in the $n = 3$ state make transitions to lower-energy states, emitting one or more photons, until finally all atoms are in the ground state. How many distinct photon energies would be observed?

24-8 Consider a very large hydrogen "atom," in which the electron orbits the proton in a circular path having a radius of 1.0 mm. What is the (approximate) quantum number characterizing the state of the atom?

24-9 A singly ionized helium atom, He^+, is hydrogenlike in that it has a single electron. What is the (a) ionization energy and (b) radius in the ground state, according to the Bohr theory, for He^+?

24-10 A beam of monoenergetic electrons, each with a kinetic energy of 12.0 eV, is used to bombard a collection of atoms in the form of a gas. It is found that most electrons emerge from the gas with a kinetic energy of 12.0 eV; some electrons, however, emerge with a kinetic energy of 8.4 eV. (a) What is the excitation energy of the atoms, i.e., the difference in energy between the first excited and the ground states? (b) What is the energy of photons one would expect to find emitted by the atoms when bombarded by electrons as given above?

WAVE PROPERTIES OF PARTICLES

CHAPTER TWENTY-FIVE

CHAPTER 25 WAVE PROPERTIES OF PARTICLES

The Bohr model was a dazzling breakthrough, but it soon revealed its shortcomings. Try as they would, physicists were unable to apply it successfully to anything more complicated than the hydrogen atom with its single electron. Even for helium the model fails to account for atomic stability: one of the two orbiting electrons should eventually drift away.

These difficulties do not necessarily mean that the key idea of the Bohr theory, stationary energy states, is wrong. Indeed, even the commonplace observation that tables and chairs do not of themselves tend to collapse or fly apart seems evidence that stationary energy states do exist. However, they need not be linked to circular orbits, in which electrons mysteriously travel without radiating. There may be a deeper explanation, one that indicates a mechanism, of why atomic energies have certain fixed values.

The present chapter will establish that not only electromagnetic waves but material particles as well exhibit dual properties. To a greater or lesser extent, every piece of matter acts like—*is*—a wave! It is this wavelike behavior that explains the stationary energy states of atomic electrons and leads to some of the great results of modern physics.

25-1 DE BROGLIE'S PREDICTION OF MATTER WAVES

In 1923 a French graduate student, Louis de Broglie, predicted that because light (which we previously thought of as a wave) has particle properties, electrons or any other material "particles" might have wave properties. He suggested a number of experiments to check his prediction, and within a few years the wave properties of particles were firmly established. We shall detail the evidence in the sections that follow.

De Broglie's basic insight is contained in his relation for the *wavelength of a particle*. Because the energy and momentum of a *photon* are related to its frequency and wavelength by $E = h\nu$ and $p = h/\lambda$, de Broglie assumed that the frequency ν and wavelength λ of a material particle are related to its energy E and momentum p in the same way, namely,

De Broglie's hypothesis: the wavelength of a particle is $\lambda = h/p$

[25-1] $$\nu = \frac{E}{h}$$

and

[25-2] $$\lambda = \frac{h}{p}$$

Here the energy of the particle is given by $E = mc^2$ and the momentum by $p = mv$.

The speed of a *wave* is found by dividing the wavelength (which is the distance traveled in one period) by the period. Using the symbol u for this wave speed (we are writing v for the particle speed), we have

$$u = \frac{\lambda}{T} = \lambda\nu = \frac{h}{p}\frac{E}{h} = \frac{\hbar}{mv}\frac{mc^2}{\hbar} = \frac{c^2}{v} \qquad [25\text{-}3]$$

For a photon, $v = c$; therefore, the wave speed u is just the speed of light. This is quite consistent with our previous understanding of light. But for material particles, which necessarily have a speed less than c, [25-3] shows that the wave speed u is greater than the speed of light. At first this may appear to conflict with the requirement of relativity that c be the ultimate speed, but actually it does not. The speed of light c is the limiting speed for the *transport of energy and momentum*, which is accomplished at the *particle speed v*, not the de Broglie wave speed u.

How could these particle waves have gone so long unnoticed? Shouldn't bullets fired through a slit show diffraction in the same fashion as light waves? The answer to this question is very simple: the wavelengths of the bullets are so small that the bullets pass through the relatively wide slit in an unbroken beam. Indeed, for a projectile having a mass of only 1 gm and moving at only 1 cm/s, the de Broglie wavelength is $\lambda = h/p = 6.6 \times 10^{-34}$ J-s/$[(10^{-3}$ kgm$)(10^{-2}$ m/s$)] = 6.6 \times 10^{-29}$ m $= 6.6 \times 10^{-19}$ Å. For this minute wavelength to produce a clear Fresnel diffraction pattern, the detector used with a slit of width 1.0 cm would have to be placed a distance D beyond the slit given by (see Examples 19-2 and 19-3) $D = W^2/8\lambda = (10^{-4}$ m$^2)/(8 \times 6 \times 10^{-29}$ m$) = 2 \times 10^{23}$ m. But the edge of the universe is believed to be at 10^{10} light years, or 10^{26} m. It would be necessary to locate the detector on another galaxy!

Ordinary objects have extremely minute wavelengths.

Any hope of observing the wave properties of large objects must be abandoned; however, de Broglie suggested using electrons as the projectiles. At low speeds their wavelength might be sufficiently long to permit detection of interference effects.

Let us determine an electron's speed after it has been accelerated through 100 V, assuming that the answer will be much less than the speed of light and therefore classical physics applies:

$$\Delta E_k = -\Delta E_p$$

or

$$\tfrac{1}{2}mv^2 = eV$$

whence

$$v = \sqrt{\frac{2eV}{m}} = \sqrt{\frac{(2)(1.6 \times 10^{-19})(100)}{9.1 \times 10^{-31}}} \approx 6 \times 10^6 \text{ m/s}$$

(the classical calculation is seen to be appropriate). The wavelength of the electron is then

$$\lambda = \frac{h}{mv} \approx \frac{6.6 \times 10^{-34}}{(9.1 \times 10^{-31})(6 \times 10^6)} \approx 0.1 \times 10^{-9} \text{ m} = 1 \text{ Å}$$

An x-ray photon and a 100-eV electron both have a wavelength of about 1 Å.

which is enormously greater than the wavelength of the bullets described above. Indeed, a 100-eV electron has a wavelength comparable to that of an x-ray, and this provides a clue to a way of observing the diffraction of electrons.

EXAMPLE 25-1 Calculate the de Broglie wavelength associated with a 0.20-kgm billiard ball moving with a speed of 0.50 m/s. Express your answer in angstroms.

SOLUTION

$$\lambda = \frac{h}{p} = \frac{6.6 \times 10^{-34} \text{ J-s}}{(0.2 \text{ kgm})(0.5 \text{ m/s})}$$

$$= (6.6 \times 10^{-33} \text{ m})(10^{10} \text{ Å/m}) = 6.6 \times 10^{-23} \text{ Å}$$

That is, the wavelength of this billiard ball is so small as to be completely undetectable.

25-2 OBSERVATION OF THE WAVELENGTHS OF X-RADIATION

For light waves, the ruled grating furnishes a powerful means for measuring wavelengths. In Section 18-9 we found that the successive major intensity maxima can be readily resolved if the grating spacing d does not too much exceed the wavelength λ. Accurate investigation of x-rays or the de Broglie waves of laboratory electrons thus calls for a grating with a spacing of only a few angstroms.

A crystal as a diffraction grating

In 1912 Max von Laue, a theoretical physicist, suggested that a crystal, with its regularly spaced atoms, might serve as a grating for x-rays. It was known that the distance between atoms was of the order of a few angstroms. Aided by Friedrich and Knipping, von Laue almost immediately succeeded in obtaining a diffraction pattern from a single crystal of rock salt, NaCl. The apparatus is shown in

+ SECTION 25-2
OBSERVATION OF
THE WAVELENGTHS
OF X-RADIATION

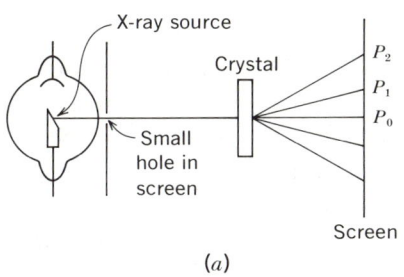

Fig. 25-1a. A typical pattern for a single crystal, given in Fig. 25-1b, consists of a central spot surrounded by symmetrically located secondary spots.

This kind of diffraction pattern was interpreted by W. L. Bragg in 1913. He assumed that the spots of maximum intensity on the screen result from perfectly constructive interference of radiation scattered from all the atoms in certain planes of the crystal. These planes, now called *Bragg planes*, are identified in Fig. 25-2. The interference, of course, is more complicated than for a line grating because the sources of radiation are not simple parallel lines or slits but individ-

FIGURE 25-1 The diffraction of x-rays by a crystal: (*a*) the experimental arrangement, with a narrow beam incident upon a crystal and the points $P_0, P_1, P_2, \ldots$ of strong intensity recorded on a distant screen; (*b*) a photograph of the pattern appearing on the screen for diffraction by a crystal of NaCl. (*From "Introduction to Physics and Chemistry" by Beiser and Krauskopf. Copyright © 1964 by McGraw-Hill, Inc.*)

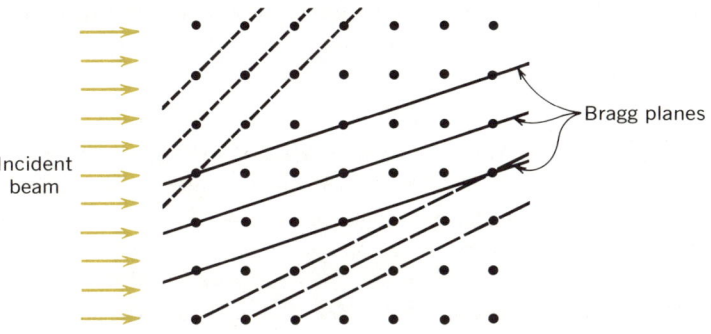

FIGURE 25-2 Three sets of Bragg planes for a crystal in which the atoms are situated in a cubical array.

CHAPTER 25
WAVE
PROPERTIES OF
PARTICLES

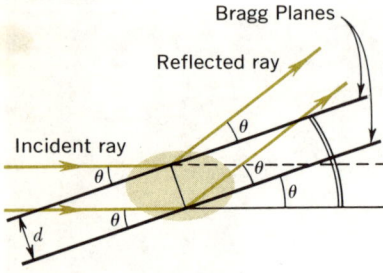

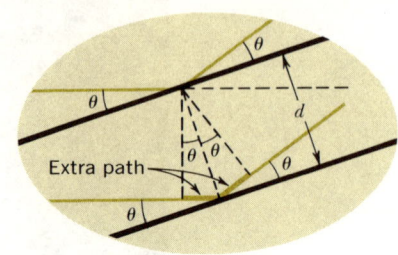

Enlarged view

FIGURE 25-3 A beam incident upon a pair of Bragg planes separated by the distance d. Both the incident and reflected rays make an angle θ with respect to the Bragg planes. The lower ray has an extra path distance of $2d \sin \theta$.

[1] This makes sense. In fact, ordinary reflection is precisely due to reradiation by the atoms composing the reflecting surface.

The Bragg relation

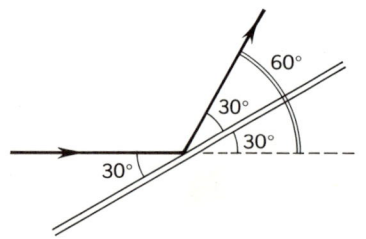

FIGURE 25-4

EXAMPLE 25-2

SOLUTION

ual atoms distributed in three dimensions. But the basic principle is the same: each point source emits an electromagnetic wave which may interfere with the waves from all the other sources.

The central bright spot (P_0 in Fig. 25-1a) comes, of course, from the undeviated transmitted radiation. But what about the other spots? Bragg supposed that each set of parallel Bragg planes reflects the incident radiation like so many parallel mirrors.[1] The angle θ in Fig. 25-3 is between the direction of the incident beam and the Bragg plane; it is the *complement* of the angle of incidence as defined in Chapter 17. From the enlarged view in the second part of Fig. 25-3 it can be seen that if the distance between two adjacent Bragg planes is d, then the radiation reflected from the second plane at an angle θ with that plane travels a distance $2d \sin \theta$ farther than the radiation reflected from the first plane at the same angle. If this additional distance is an integral number of wavelengths, i.e., if

$$2d \sin \theta = n\lambda \qquad [25\text{-}4]$$

there will be perfectly constructive interference at angle θ with the Bragg planes or at angle 2θ with the incident radiation.

If the spot P_1 in Fig. 25-1a corresponds to first-order Bragg reflection ($n = 1$), then P_2 might correspond to second-order Bragg reflection ($n = 2$) (or the two spots might correspond to reflections from different sets of Bragg planes).

The first-order Bragg reflection from a single crystal of NaCl (crystal spacing $d = 2.82$ Å) occurs at an angle of 60° with respect to the direction of the incident beam (see Fig. 25-4). What is the wavelength of the incident x-rays?

When the angle between the reflected beam and the incident direction is 60°, the angle between the reflected beam and the reflecting planes is 30°. From [25-4] we have

$$\lambda = \frac{2d \sin \theta}{n} = \frac{2(2.82 \text{ Å})(\sin 30°)}{1} = 2.82 \text{ Å}$$

EXAMPLE 25-3 Show that there is no possibility of perfectly constructive interference if the wavelength of the incident radiation and the crystal spacing are such that $2d$ is less than λ.

SOLUTION From Eq. [25-4]

$$\sin \theta = n \frac{\lambda}{2d}$$

If $2d < \lambda$, $\sin \theta$ would be greater than 1 (recall that $n \geq 1$), but there is no θ which satisfies this condition. The physical meaning is easy to see. Consider the simplest case, first-order reflection ($n = 1$) and maximum path difference ($\theta = 90°$, or backward reflection) (see Fig. 25-5). In order to obtain perfectly constructive interference we must at least require that $2d = \lambda$; for if $2d$ is less than λ, the disturbance reflected from the second plane can never be identical to that from the first plane.

The diffraction pattern shown in Fig. 25-1b for a single crystal consists of spots because the cross section of the incident x-ray beam is small and is thus itself a spot and because this incident spot is reflected by finitely many sets of Bragg planes which are well separated from each other in angle. However, another type of diffraction pattern is obtained when a *powdered sample* of small crystal flakes is used. In one such pattern, illustrated in Fig. 25-6b, the intensity maxima are concentric *rings*. This is just what one expects. In a powder the Bragg planes have a continuous distribution over all possible orientations, because the many small crystals are randomly oriented. For each set of planes the Bragg condition [25-4] still determines the angle 2θ of maximum reflected intensity, but now the whole pattern must be rotationally symmetric about the incident beam. Thus, each order of reflection generates a cone in space, which is intercepted as a ring on the screen (Fig. 25-6a).

Crystals, then, make ideal instruments for measuring wavelengths in the x-ray range. Conversely, by using incident radiation of known wavelength, we can infer crystal structure from the diffraction pattern. In particular, from the relative intensities of the spots or rings

SECTION 25-2
OBSERVATION OF
THE WAVELENGTHS
OF X-RADIATION

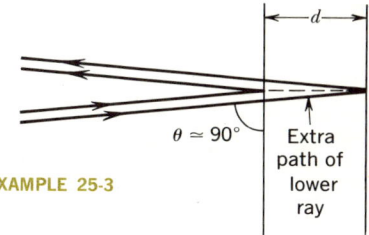

FIGURE 25-5

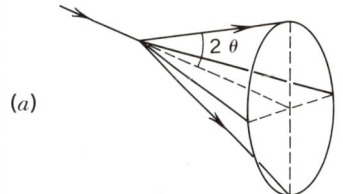

(a)

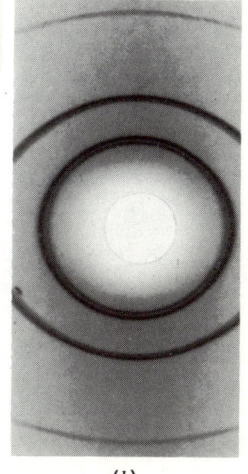

(b)

FIGURE 25-6 Bragg diffraction from a powdered sample: (a) the Bragg condition for strong reflection at the angle 2θ generates a cone in space; (b) the cones for various Bragg planes are intercepted as a series of concentric diffraction rings on the screen. (*From "Elements of X-ray Crystallography" by Azaroff. Copyright © 1968 by McGraw-Hill, Inc.*)

we can gauge the relative atomic densities in the various Bragg planes. The greater the concentration of atoms in a plane, the more radiation it will reflect.

25-3 THE OBSERVATION OF MATTER WAVES

De Broglie himself suggested that either line gratings or crystals might be used to demonstrate the wave properties of electrons, but before the experiments had been completed, a curious thing happened. In 1926 two American physicists, C. J. Davisson and L. H. Germer, found some peculiar results in the classical electron-scattering experiments they were doing. At this time neither man had heard about de Broglie's hypothesis. At a conference in England, Davisson showed some of his preliminary data to several European physicists, and in the conversation that followed the suggestion was made that these strange results might be related to electron *diffraction*. Davisson and Germer immediately began to check this suggestion, and by 1927 they had used a crystal grating to establish beyond reasonable doubt that electrons can be diffracted and that their wavelength is precisely as predicted by de Broglie. Davisson and Germer studied the intensity pattern of electrons reflected from the surface of a pure crystal, but in 1928 G. P. Thomson (the son of J. J. Thomson) observed diffraction patterns from narrow electron beams passing through thin metal foils. These intensity patterns exhibited all the features of the x-ray patterns discussed in the previous section—intensities which vary with the atomic density of the Bragg planes, discrete spots for single crystals, and concentric rings for polycrystalline foils (both in Fig. 25-7). Diffraction patterns have been observed at the edge of an electron beam (Fig. 25-8, p. 576). The results again confirmed de Broglie's hypothesis: a wavelength $\lambda = h/p$ must be associated with a particle having momentum p.

Electrons reveal wave properties in diffraction.

25-4 THE HEISENBERG UNCERTAINTY PRINCIPLE

The discovery that objects such as electrons have wave properties closed the circle between wave and particle. Evidently, nothing exists purely as a wave or purely as a particle but partakes of a dual nature. A basic implication of this two-sidedness is contained in *Heisenberg's uncertainty principle*. It says that there are some questions about physical phenomena which can be answered with just so much exactness and no more—not because our measuring instruments are crude but because of the very workings of the universe. More precisely, Heisenberg's principle states that the quantities involved in the description of a physical event fall into certain pairs and that it is impossible

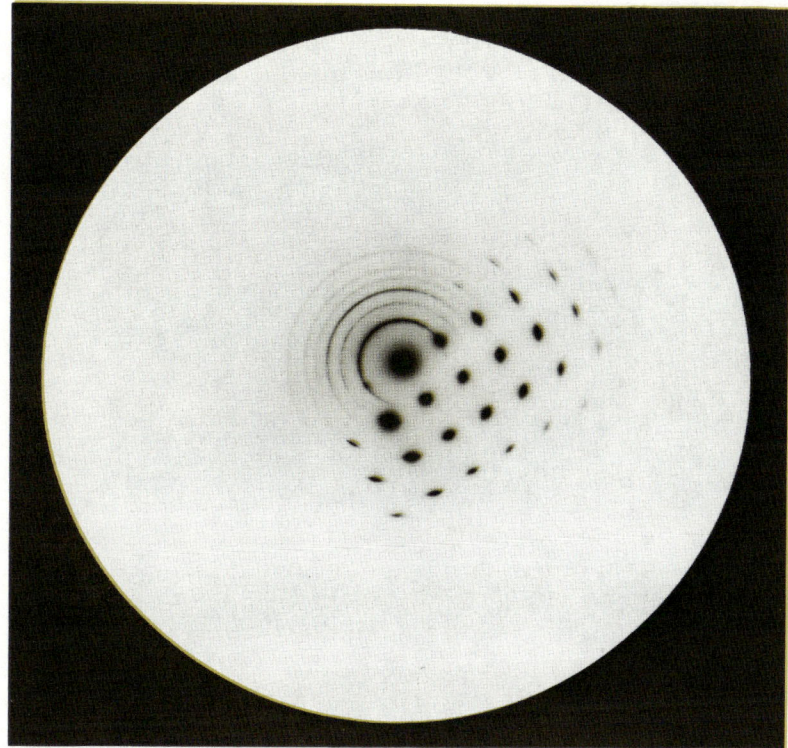

FIGURE 25-7 Electron diffraction from a thin film which was partly a single crystal (*lower right*) and partly polycrystalline (*upper left*). (*Sigrid Herd, IBM Corporation.*)

at the same time to specify both members of a given pair to arbitrarily high precision.

Let us see how the uncertainty principle works for the pair of quantities position and momentum. In Fig. 25-9 a simple diffraction experiment is illustrated. A very wide beam of electrons (photons would do as well) is incident normally on a plane containing a single narrow slit of width w. Some of the electrons pass through the slit, and their arrival at a distant screen is observed. Especially if the incident beam is weak, these arrivals can be detected one at a time. For only a few electrons, it is difficult to make out any pattern in the positions registered on the screen, but after a large number have arrived, the pattern shown in Fig. 25-9 is clearly defined. It is the standard single-slit Fraunhofer diffraction pattern (see Fig. 19-16c).

What does this mean in terms of our particle description of the electrons? Somehow the electrons which pass through the slit are deflected from the original direction of their motion. Deflected by how much? No one has ever been able to find a way to predict the

576
CHAPTER 25
WAVE
PROPERTIES OF
PARTICLES

FIGURE 25-8 Photograph of the diffraction of a beam of electrons at a straight edge. (*From H. Raether, "Elektroninterferenzen," in "Handbuch der Physik," XXXII, Springer-Verlag, Berlin-Göttingen-Heidelberg, 1957.*) Compare with Figure 19-9.

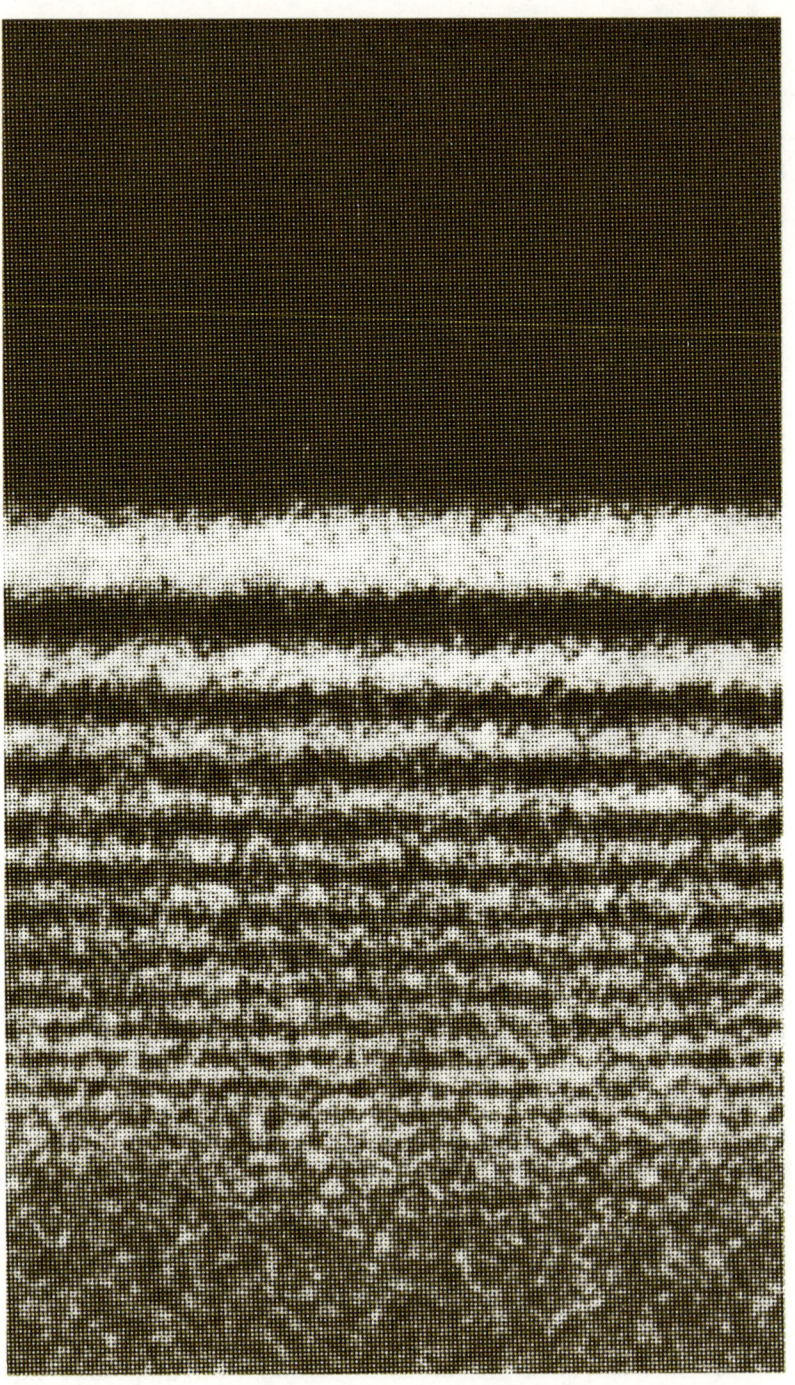

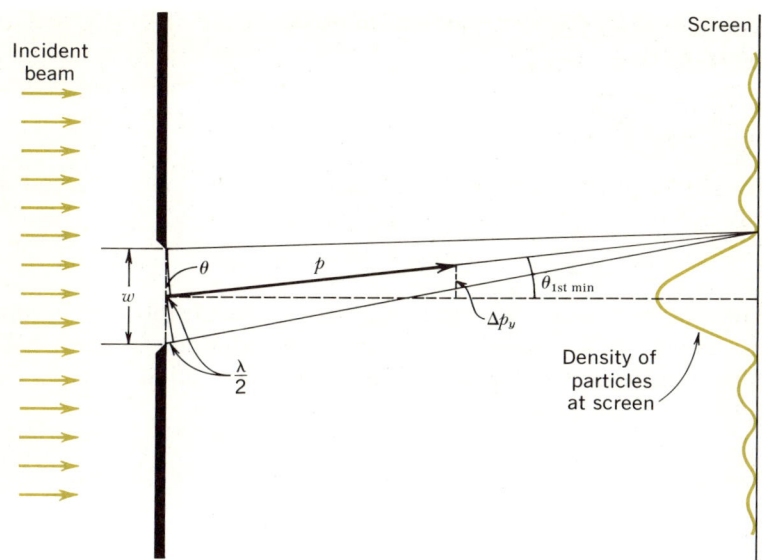

FIGURE 25-9 A beam of electrons is incident upon a slit of width w, and a diffraction pattern appears on the distant screen. An electron arriving at the screen near the location $\theta_{\text{1st min}}$ of the first minimum in the diffraction pattern has a y momentum component of Δp_y.

deflection of a *particular* electron; but with respect to a *large number* of electrons, Fig. 25-9 tells us that most of them will land within the angular deflection $\theta_{\text{1st min}}$, which is associated with the minimum in intensity nearest the center of the pattern. According to the figure, the maximum y momentum an electron can have and still hit within the central bright fringe is given by Δp_y, where

$$\sin \theta_{\text{1st min}} \approx \Delta p_y / p \qquad [25\text{-}5]$$

Only about 15 percent of the electrons are found to have transverse momenta of magnitude greater than Δp_y; the remainder have p_y magnitudes that are distributed between 0 and Δp_y. The Δp_y so defined is a good measure of the uncertainty in the y momentum of the electrons emerging from the slit.

Before the electrons passed through the slit, there was no uncertainty in their momentum; they were all moving normally to the plane of the slit, and p_y was zero by definition. But as soon as they go through the slit, i.e., as soon as their y positions are specified to within an uncertainty $\Delta y = w$, we find that the y momentum is no longer specified precisely but with an uncertainty Δp_y.

We have previously seen that the single-slit diffraction pattern can be explained only in terms of wave properties. It is precisely this same explanation which now allows us to relate the uncertainties Δy and Δp_y quantitatively. According to Section 19-4 and Fig. 19-14,

The Heisenberg uncertainty principle

CHAPTER 25
WAVE PROPERTIES OF PARTICLES

the angular position of the first minimum in the intensity pattern is given by

$$[25\text{-}6] \quad \sin\theta_{1\text{st min}} = \frac{\lambda/2}{w/2} = \frac{\lambda}{\Delta y}$$

By use of de Broglie's relation $\lambda = h/p$, [25-6] can be written as

$$[25\text{-}7] \quad \sin\theta_{1\text{st min}} = \frac{h}{p\,\Delta y}$$

We now have two expressions for the angular position of the first minimum: [25-5] coming from a particle description of the single-slit diffraction and [25-7] originating in a wave description. By equating these two expressions, we find

Uncertainties in position and momentum

$$\frac{\Delta p_y}{p} \approx \frac{h}{p\,\Delta y}$$

or

$$[25\text{-}8] \quad \Delta y\,\Delta p_y \approx h$$

Here is Planck's constant again, turning up in a remarkable way! The uncertainty in position Δy and the uncertainty in momentum Δp_y do have the kind of interrelation we predicted: as one is reduced, the other *inevitably* is increased, and, quantitatively, the product of these two uncertainties is of the order of $h = 6 \times 10^{-34}$ J-s. Planck's constant identifies a fundamental limitation on the simultaneous description of an object's position and momentum.

Like the wave nature of matter itself, the uncertainty principle plays an important role only in the microscopic domain. If, for example, instead of electrons we considered a stream of bullets passing through an ordinary window of width $\Delta y = 1$ m, the uncertainty in momentum of each bullet would be about $\Delta p_y \approx h/\Delta y = 6.6 \times 10^{-34}$ kgm-m/s, which represents a negligible fraction of the whole momentum. This result is consistent with what we found in Section 25-1: the bullets have so short a wavelength that they are essentially unaffected by the window.

EXAMPLE 25-4   In Fig. 25-9 the incident electrons at the far left move in a direction *precisely* normal to the plane of the slit. There is an uncertainty in p_y only to the right of the slit; to the left of the slit $\Delta p_y = 0$. Is this inconsistent with Heisenberg's uncertainty principle?

SOLUTION No; because the width of the incident beam is not limited, we must take Δy to be infinite to the left of the slit. Then $\Delta y \, \Delta p_y$, being the product of infinity and zero, is indeterminate, which does not contradict $\Delta y \, \Delta p_y \approx h$.

We have described the uncertainties in an object's position and momentum components (y and p_y) perpendicular to the direction of its motion. Now consider the components (x and p_x) parallel to its motion. Supposing x can be determined to within an uncertainty Δx, we ask whether p_x can be simultaneously measured with as high a precision as we please.

Again we go over to the wave description. Knowing the longitudinal position of the object to within Δx means knowing that its wave vanishes outside an interval of length Δx along the direction of propagation. In other words, the wave is assumed to be a pulse of finite length Δx. In this situation, measuring the object's momentum with such and such precision is, by de Broglie's formula, the same as measuring the wavelength within this pulse with a certain precision.

The question, then, reduces to whether it is possible to measure the wavelength within finite pulses with unlimited precision. In Section 18-9, where we used a grating to measure wavelength, we found that the larger the number of slits, the narrower the fringes (intensity maxima) and hence the greater precision with which the wavelength can be measured. One might now think of using a grating with arbitrarily many slits ($N \to \infty$) to obtain unlimited precision. Let us see whether this idea will work, starting with a grating for which $N = 6$ (Fig. 25-10). Consider the first-order intensity maximum. The wavelet from slit 2 travels exactly 1 wavelength farther in reaching the position of this maximum than the wavelet from slit 1 does; that from slit 3, exactly 2 wavelengths, etc.; but as we saw in Section 18-9, the crucial condition for obtaining very narrow intensity maxima, and hence very precise wavelength measurements, is that light from the upper $N/2$ slits interfere perfectly destructively with light from the lower $N/2$ slits at a point on the screen only a small distance away from the principal intensity maximum (see Figs. 18-25c and 18-26).

For this condition to be satisfied, even for as short a time as one period, the length Δx of the incident pulse must be at least equal to $N\lambda$. Why? Look again at Fig. 25-10 and imagine what would

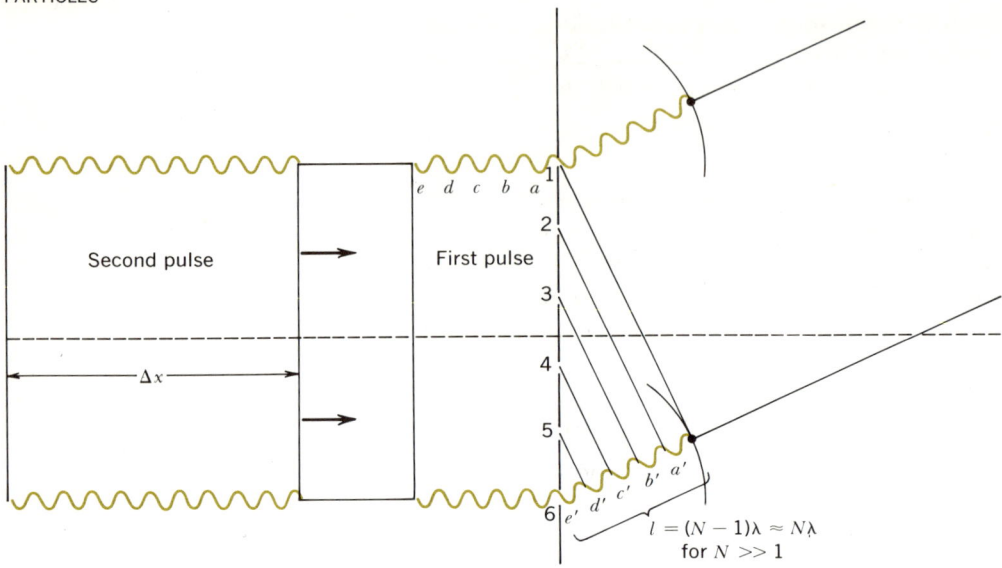

FIGURE 25-10 Pulses each of length Δx incident upon a grating with six slits. The waves emerging from slits 1 and 6 are shown in the direction toward the first-order intensity maximum.

happen if Δx equalled λ instead of 10λ. The entire wave disturbance from slit 1 would arrive at the position on the screen where the principal maximum is expected and would *be completed before the signal from slit 2 had arrived.* The situation would be even worse for the signals from slits 3, 4, 5, and 6. No interference effects at all would be possible, and a sharp principal maximum would be out of the question. It is not until $\Delta x = 6\lambda$ (the grating in Fig. 25-10 has $N = 6$ slits) that there will be wave disturbances arriving simultaneously at the position of the principal maximum from *all* slits. Thus, unless

[25-9] $\quad \Delta x \geq N\lambda$

i.e., unless $N \leq \Delta x/\lambda$, the grating is useless for precise wavelength measurements. This spells doom for our idea of getting infinite precision by letting N approach infinity so long as the length of the pulse remains Δx.

The conclusion is that the wavelength of a finite pulse can be measured only with limited precision. Translated back into particle language, this means that an uncertainty in momentum Δp_x accompanies the position uncertainty Δx. As the detailed analysis in Example 25-7 will show, the two uncertainties are related by

[25-10] $\quad \Delta x \, \Delta p_x \approx h$

which is precisely the same form the Heisenberg uncertainty principle takes for the transverse components y and p_y.

The results of the two special problems we have just considered, [25-8] and [25-10], are found to be universally applicable. In both examples we extracted the utmost accuracy possible. Therefore, h represents a minimal estimate for the product of the uncertainties, and in general we must write

$$\Delta x \, \Delta p_x \gtrsim h \qquad \Delta y \, \Delta p_y \gtrsim h \qquad \Delta z \, \Delta p_z \gtrsim h \qquad [25\text{-}11]$$

where the symbol $\gtrsim$ is read "greater than or approximately equal to."

Two other important physical quantities are jointly subject to the uncertainty relation. For the *energy* E of an object and the *time* t for which this energy is possessed, we have (see Problem 25-10)

$$\Delta E \, \Delta t \gtrsim h \qquad [25\text{-}12]$$

Query For a photon the energy-time relation follows directly from the coordinate-momentum relation. Can you show how?

EXAMPLE 25-5

Consider an electron confined to a volume having dimensions comparable to the diameter of a hydrogen atom, i.e., approximately 1 Å, or 10^{-10} m. Calculate the minimum uncertainty in its speed and the minimum value of its average kinetic energy. Compare this minimum energy with the potential energy associated with the electrostatic attraction between an electron and a proton approximately $\frac{1}{2}$ Å apart.

SOLUTION

The uncertainty in any component of the electron's velocity is given by Heisenberg's uncertainty principle [25-11]

$$\Delta x \, (m \, \Delta v_x) \gtrsim h$$

or

$$\Delta v_x \gtrsim \frac{h}{m \, \Delta x} = \frac{6.6 \times 10^{-34}}{9.1 \times 10^{-31} \times 10^{-10}} \approx 7 \times 10^6 \text{ m/s}$$

Note that the minimum *uncertainty* in speed, 7×10^6 m/s, is larger than the actual speed of the electron according to either the classical planetary model or the Bohr model (Problem 24-4). Clearly, no satisfactory analysis of the hydrogen atom can ignore the wave properties of the electron.

Now for the minimum kinetic energy of the electron, which can be computed from the minimum speed. The minimum average speed must be at least of the order of one-half the uncertainty in the speed.

To see why, think of an airplane whose total fluctuation (uncertainty) in altitude is given as 1,000 ft. This means that the airplane may sometimes be as much as 500 ft above and sometimes as much as 500 ft below its average cruising altitude. Certainly, then, the plane cannot be cruising below 500 ft.

Thus, for an electron in the hydrogen atom the Heisenberg uncertainty principle implies that the *minimum* average speed is about 3.5×10^6 m/s. Because this is a nonrelativistic speed, the minimum average kinetic energy of the electron is given by

$$(E_k)_{min} = \tfrac{1}{2}mv_{min}^2 \approx \tfrac{1}{2}(9.1 \times 10^{-31})(3.5 \times 10^6)^2$$
$$\approx 50 \times 10^{-19} \text{ J} \approx 30 \text{ eV}$$

Our earlier calculations using the Bohr model show that a purely electrostatic interaction would give a potential energy of about -27 eV. The agreement in magnitude is close enough to suggest that we are still justified in assuming that electrostatic attraction accounts for the binding of an electron and proton in a hydrogen *atom*. Not so in the *nuclear* problem considered in the next example.

EXAMPLE 25-6 It was once believed that nuclei might contain very tightly bound electron-proton pairs. Such pairs, which would have no net charge and a mass approximately equal to that of a proton, might account for the known relationships between the charge numbers and mass numbers of nuclei.

A typical nuclear dimension is 10^{-15} m. What would be the uncertainty in the momentum p for the electrons in these pairs? What would be the minimum average kinetic energy of such electrons? Compare this minimum energy with the potential energy associated with the electrostatic attraction between an electron and proton separated by 10^{-15} m.

SOLUTION Again the Heisenberg uncertainty principle can be used to determine the minimum uncertainty Δp in the magnitude of the momentum p

$$\Delta p_x \gtrsim \frac{h}{\Delta x} = \frac{6.6 \times 10^{-34}}{10^{-15}} = 6.6 \times 10^{-19} \text{ kgm-m/s} \approx 10^3 \frac{\text{MeV}}{c}$$

As you should verify, a momentum of the order of 10^{-19} kgm-m/s implies a relativistic speed for the electron. Applying the relativistic energy formula [22-22] with $p \approx 500$ MeV/c, we find that the electronic rest energy (0.5 MeV) is quite negligible in comparison with its kinetic energy $E_k = E - E_0$

$E_k \approx E \approx pc \approx 500$ MeV

The electrostatic potential energy for a proton-electron pair varies inversely as the separation; and we know that for the atom-sized separation 0.5 Å $= 5 \times 10^{-11}$ m, the value is -27 eV. Thus, for a separation of 10^{-15} m

$E_p \approx \dfrac{5 \times 10^{-11}}{10^{-15}}(-27 \times 10^{-6} \text{ MeV}) \approx -1$ MeV

With only 1 MeV of binding energy available, it is clear that electrostatic attraction could never confine a 500-MeV electron inside the nucleus. Indeed, electron-proton pairs cannot exist in the nucleus; then how can one account for the nuclear mass and charge? This central question of nuclear physics will be considered in Section 27-2.

Electrons cannot exist in an atomic nucleus.

Show that the Heisenberg uncertainty relation emerges from an analysis of a wave having limited extension Δx along the direction of its propagation and impinging upon a grating, as shown in Fig. 25-10.

+ EXAMPLE 25-7

First we need an estimate of the precision with which the wavelength within the wave pulse can be measured using a grating. This is particularly easy when the angle θ to the principal intensity maximum in the first order is small. In Section 18-9 we found that the *angular* position of the *first* principal intensity maximum ($n = 1$) is given by

SOLUTION

$\theta = \dfrac{\lambda}{d}$

[25-13]

where d is the distance between adjacent slits. The angular displacement of the nearest intensity minimum from a principal intensity maximum is λ/Nd (see [18-11]), where N is the number of slits in the grating.

Of course, when we look at a light fringe we cannot tell precisely where its center is. But we can estimate that the center lies somewhere in a total spread $\Delta\theta$ of angular positions around the position θ. We can write this uncertainty in the position of the intensity maximum as $\theta \pm \tfrac{1}{2}\Delta\theta$. Similarly, there will be an uncertainty in the wavelength we calculate using [25-13]; this uncertainty can be indicated by writing the calculated wavelength as $\lambda \pm \tfrac{1}{2}\Delta\lambda$.

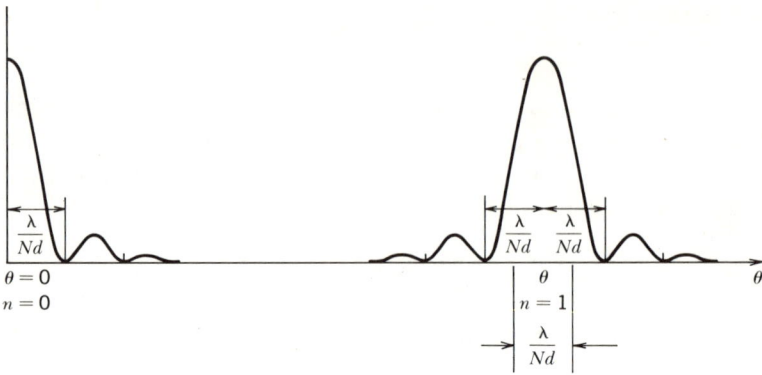

FIGURE 25-11

Therefore, [25-13] becomes

[25-14] $$\theta \pm \tfrac{1}{2}\Delta\theta = \frac{\lambda \pm \tfrac{1}{2}\Delta\lambda}{d}.$$

If we use λ/Nd as a estimate of the uncertainty in our measurement of θ [this is an angular spread covering half the angular displacement between the nearest minima, one each on either side of the principal maximum (see Fig. 25-11)], then [25-14] becomes

[25-15] $$\theta \pm \frac{1}{2}\frac{\lambda}{Nd} = \frac{\lambda \pm \tfrac{1}{2}\Delta\lambda}{d}$$

Using [25-13], we obtain

[25-16] $$\frac{\Delta\lambda}{\lambda} = \frac{1}{N}$$

Thus, at best, the grating gives a fractional uncertainty in wavelength equal to the reciprocal of the number of slits. (This applies for the first-order maximum; for the nth-order maximum the result is $\Delta\lambda/\lambda \approx 1/nN$.) For six slits the error is about 16 percent; for 1,000 slits, it is 0.1 percent.

Finally, knowing the uncertainty $\Delta\lambda$ in the wavelength measurement, we can compute the uncertainty Δp_x in the x component of momentum for our wave object. It is given by

$$\Delta p_x = \frac{h}{\lambda} - \frac{h}{\lambda + \Delta\lambda} = \frac{h\,\Delta\lambda}{\lambda^2 + \lambda\,\Delta\lambda}$$

but if $\Delta\lambda \ll \lambda$, we can write

$$\Delta p_x \approx \frac{h\,\Delta\lambda}{\lambda^2} \qquad [25\text{-}17]$$

Now if the position of our object is specified to within an uncertainty Δx, the right grating to use is one with $N = \Delta x/\lambda$ slits (see [25-9]). The precision of this grating will be $\Delta\lambda/\lambda = 1/N = \lambda/\Delta x$ (see [25-16]), or $\Delta\lambda/\lambda^2 = 1/\Delta x$. Thus, the uncertainty in momentum [25-17] becomes

$$\Delta p_x \approx \frac{h}{\Delta x}$$

or

$$\Delta x\,\Delta p_x \approx h$$

which is the Heisenberg relation.

25-5 PRINCIPLE OF COMPLEMENTARITY

There is no escaping the fact that electrons (and other material objects) display the same duality as electromagnetic radiation. Under appropriate circumstances, e.g., when a beam of electrons is sent through a crystal, we see diffraction effects. That is, our observations make sense only if we think of the electrons as waves. At the same time we recognize that electrons which first interact with the deflecting plates of a cathode-ray oscilloscope tube and then strike the screen, causing tiny bright flashes, have to be regarded as particles.

What is an electron *really?* The question is pressed on us by our discomfort in thinking of it as simultaneously a wave and a particle. Further reflection shows that the very concepts of particle and of wave are contradictory and mutually exclusive. A particle is, by definition, an object whose internal structure can be ignored. Therefore its mass, momentum, and energy can be assigned to a sharply localized region of space. A wave is quite another thing. To understand its interference and diffraction effects, we must consider its internal structure. Moreover, if we are to measure a *precisely* defined wavelength or frequency, the wave must extend over a large distance in space. We have already shown in Example 25-7, using strictly non-quantum-mechanical arguments, that a wave whose extension in space along the direction of propagation is limited has a wavelength that can be determined only with uncertainty. Indeed, if it were to be completely localized, we could not assign any wavelength to it whatsoever. But a particle has a high degree of localizability, just the opposite of a wave. It is really impossible to think *simultaneously* of any disturbance as both an ideal particle and a wave.

Bohr's principle of complementarity says that information about the wave and particle aspects of objects, obtained under different experimental conditions, represents equally essential knowledge about the object and together these aspects constitute the complete knowledge. Moreover, this information about a realm of experience remote from ordinary life cannot be combined into a *single* picture by means of ordinary concepts. This is not surprising because in no *single* experiment does an object show all its properties; either the wavelike is displayed and the particlelike suppressed or vice versa. Although matter and radiation possess both wave and particle characteristics, they never exhibit both simultaneously.

This "either or" of Bohr's principle rationalizes the quantitative "not both" of the uncertainty relation. Wave language and particle language complement each other perfectly; together, but not at the same time, they can provide complete understanding. Total reliance on either one is inadequate, and the simultaneous use of both leads to the frustration described by the uncertainty conditions. So it was, for instance, with our investigation of a finite pulse: the more precisely we could locate "the particle," the less precisely we could measure "the wavelength."

The complementarity principle: different experimental conditions demand different but equally essential concepts (e.g., wave and particle).

25-6 THE PROBABILITY INTERPRETATION OF MATTER WAVES

When such hallmarks of waviness as interference and diffraction are exhibited by x-rays or gamma rays, we know that what waves is an electromagnetic field. But when an electron or (imperceptibly) a golf ball is diffracted, what waves? To answer this question let us reexamine the case of electromagnetic waves.

The electric field **E** at any point is primarily defined as the electric force per unit positive charge located at that point. However, in Section 13-7 we developed another interpretation. Since energy is required to establish an electric field in space, energy in the same amount can be considered as having been stored in the field itself. The magnitude of **E** then appears as a measure of the volume concentration of this energy, according to [13-8]

[25-18] $\quad$ Electric energy density $= u_e = \dfrac{E^2}{8\pi k_e}$

Energy is also associated with a magnetic field **B**. It can be shown that the magnetic energy per unit volume u_b is given by a relation completely analogous to [25-18]

[25-19] $\quad$ Magnetic energy density $= u_m = \dfrac{B^2}{8\pi k_m}$

Recall that the fundamental electric and magnetic constants, k_e and k_m, are related to the speed of light by $c = \sqrt{k_e/k_m}$. Moreover, in an electromagnetic wave, the magnitudes of **E** and **B** obey $B = E/c$. Therefore we can write [25-19] as

$$u_m = \frac{B^2}{8\pi k_m} = \frac{(Bc)^2}{8\pi k_e} = \frac{E^2}{8\pi k_e} = u_e$$

For an electromagnetic wave, the total energy per unit volume is then

$$u = 2u_e = \frac{E^2}{4\pi k_e} \qquad [25\text{-}20]$$

In short, the energy density of an electromagnetic wave at any location is proportional to the square of the electric field there.

Quantum theory looks at the situation quite differently. If a beam consists of photons of a single frequency, the energy of the beam per unit volume is simply the energy $h\nu$ of any one photon multiplied by the total number of photons per unit volume. Suppose, then, that the electric field of an electromagnetic wave at some point in space has the magnitude $E = 11$ N/C. From [25-20] the corresponding electromagnetic energy per unit volume is

$$u = \frac{E^2}{4\pi k_e} = \frac{(11 \text{ N/C})^2}{4\pi(9.0 \times 10^9 \text{ N-m}^2/\text{C}^2)} = 1.1 \times 10^{-9} \text{ J/m}^3$$

Suppose, further, that the beam consists of photons of visible light, each with an energy of 4.0 eV; then the number n of photons per unit volume is

$$n = \frac{u}{h\nu} = \frac{1.1 \times 10^{-9} \text{ J/m}^3}{(4.0 \text{ eV})(1.6 \times 10^{-19} \text{ J/eV})}$$
$$= 1.7 \times 10^9 \text{ photons/m}^3 = 1.7 \text{ photons/mm}^3$$

Of course, there is no such thing as seven-tenths of a photon. One observes either whole photons or none at all. To say that there are 1.7 photons per cubic millimeter implies that this is the *average* number of photons one would find in making repeated measurements of the number in 1 mm³. At one instant one might find two photons; at another, none; and so on. Therefore, the energy density u, as computed from the beam's electric field, measures not the exact number of photons found at any instant but the number one is most likely to find with repeated trials. To put it another way, *the quantity that is waving in an electromagnetic wave, the electric field, when squared is proportional to the probability of finding a photon.*

Probability of finding a photon: electric field squared

TABLE 25-1

	Photons	Material Objects
Wave type	Electromagnetic	De Broglie $\left(\lambda = \dfrac{h}{mv}\right)$
Wave function	E	ψ
Probability of finding particle	$\propto E^2$	$\propto \psi^2$
Speed of wave	c	c^2/v

An exactly parallel relation applies to a material particle. That is, the wave property, or *wave function* ψ, associated with such an object must be connected with the probability of finding the object. Indeed, the value at any point in space of the wave function ψ, when squared, is proportional to the probability of finding the material object at that location.[1]

Probability of finding a particle: wave function squared

[1] This interpretation of ψ^2 as a probability is due to Max Born (1882–1970).

The analogous relations that hold between the electromagnetic field and the particles (photons) of this field and between the wave function ψ and material particles are summarized in Table 25-1.

It was pointed out in Section 25-3 that the de Broglie wave of a material particle cannot represent the transport of energy or momentum, since its speed c^2/v exceeds the speed of light. Therefore, it is physically unobservable. This is consistent with our new understanding of a de Broglie wave as a pure probability carrier, of which the square has physical meaning.

SUMMARY

Like photons (radiation), material objects have both particlelike and wavelike properties. For an object of energy E and momentum p, the frequency and the de Broglie wavelength are

[25-1] $\quad \nu = \dfrac{E}{h}$

[25-2] $\quad \lambda = \dfrac{h}{p}$

If the position of an object is known to within an uncertainty Δx, then the limitation on the precision with which its momentum can be simultaneously known is given by the Heisenberg uncertainty relation

[25-11] $\quad \Delta x \, \Delta p_x \gtrsim h$

A similar uncertainty relation exists between the energy of a system (or object) in a certain state and the time the system stays in that state

$$\Delta E \, \Delta t \gtrsim h \qquad [25\text{-}12]$$

The square of the wave property, the wave function ψ, associated with a material object at a given location is proportional to the probability of finding the object at that location.

PROBLEMS

25-1 (a) A 5.0-gm bullet travels at 200 m/s. What is its wavelength? (b) At what speed would its wavelength be comparable to that of visible light (about 5000 Å)?

25-2 At what speed does an electron have a wavelength equal to that of visible light of 5000-Å wavelength?

25-3 (a) What is the average kinetic energy in electron volts of a neutron in thermal equilibrium with other objects at room temperature? (Such a neutron is called a *thermal neutron*.) (b) What is the wavelength of a thermal neutron (neutron mass = 1.67×10^{-27} kgm; $k = 1.38 \times 10^{-23}$ J/°K)?

25-4 What is the angle subtended by the central diffraction maximum (as viewed from the opening) of visible light (5000 Å) passing through a slit having the width of (a) a relatively large molecule (10 Å), (b) a pinhole (0.1 mm), and (c) the Palomar telescope opening (200 in.)?

25-5 Electrons from a heated filament are accelerated through an electric potential difference of 10,000 V between the filament and the anode of the electron gun. (a) What is the wavelength of these electrons? (b) A narrow beam of electrons from a small hole in the anode then passes through a thin sheet of aluminum (distance between two adjacent Bragg planes is 4.05 Å). What is the angle of deviation for the first-order diffraction pattern?

25-6 If a proton is confined to a volume having diameter 1×10^{-14} m, what is (a) the minimum uncertainty in its speed and (b) the minimum value of its average kinetic energy? (c) How large a negative charge a distance 10^{-14} m from the proton would be required to account for this amount of energy?

25-7 Show that the wavelength λ of a particle is (a) inversely proportional to $\sqrt{E_k}$, the square root of the particle's kinetic energy, when $E_k \ll E_0$, its rest energy, but (b) inversely proportional to E_k when $E_k \gg E_0$.

25-8 The lines emitted in the spectrum from atoms are not perfectly sharp. One reason why the lines have a finite width is that the quantized energy levels of the atom are not perfectly sharp; this, in turn, is due to the fact that an atom in an excited state does not remain in that state for an infinite period of time but makes a spontaneous downward transition with the emission of a photon. Uncertainty in the period of time an atom will remain in an excited state implies, through the uncertainty principle, a corresponding uncertainty in the energy of that excited state. Suppose that the average time a large collection of atoms will remain in some one particular excited state before making a transition with the emission of a photon is 10^{-8} s. (a) What is the approximate uncertainty in the energy (in electron volts) of the excited state? (b) What is the fractional uncertainty in the energy of the 5.0-eV photons emitted from such an excited state?

25-9 (a) A certain billiard ball of 100 gm mass is moving east at 10 cm/s. Its location along an east-west line is uncertain by 0.1 mm. What is the uncertainty in the billiard ball's speed? (b) Repeat this calculation for an electron ($m = 9.1 \times 10^{-31}$ kgm) traveling east at 10^4 m/s and with an uncertainty in its location of 1.0 Å.

c 25-10 Derive the Heisenberg uncertainty principle written in the form $\Delta E\, \Delta t \gtrsim h$ from the uncertainty relation written as $\Delta x\, \Delta p_x \gtrsim h$. *Hint:* From the (classical) relation between energy and momentum, $E = p^2/2m$, find by taking the differential of both sides of the equation how an energy uncertainty ΔE is related to a momentum uncertainty Δp. Eliminate Δp to find ΔE in terms of the uncertainty in position Δx. Relate the uncertainty in the position of a particle Δx traveling at speed v to the uncertainty in the time Δt it is observed within that spatial interval.

25-11 At some one instant the magnitude of the electric field of an electromagnetic wave is 4×10^{-4} N/C at location A and 1×10^{-4} N/C at location B. What is the ratio of the average number of photons at location A to that at location B?

25-12 An electron-diffraction pattern appears on a distant screen when a beam of monoenergetic electrons passes through a single slit. The number of electrons arriving at three different locations along the screen in the same time interval is: point A, 1×10^6; point B, 4×10^6; and point C, 16×10^6. What is the ratio of the electron wave function (a) at point B relative to point A and (b) at point C relative to point A?

QUANTUM SYSTEMS OF PARTICLES

CHAPTER TWENTY-SIX

"In this communication I wish to show, first for the simplest case of the ... hydrogen atom, that the usual rules of quantization can be replaced by another postulate, in which there occurs no mention of whole numbers. Instead, the introduction of integers arises in the same natural way as, for example, in a vibrating string, for which the number of nodes is integral. The new conception can be generalized, and I believe that it penetrates deeply into the true nature of the quantum rules."

So begins Erwin Schrödinger's famous paper of 1926, in which the *wave equation* for material particles was introduced. Our discussion of matter waves and the uncertainty principle in Chapter 25 showed that the Bohr model of hydrogen, with its sharply defined planetary orbits for the electron, cannot be taken as a fundamental description. It leaves out the wave nature of the atomic particles, and that is the reason it yields little more than the correct energy levels for the simplest of atomic systems. Schrödinger's approach, on the other hand, puts wavelike behavior at the very foundation. His postulated equation arises from an analogy between the motion of particles and the motion of light waves in the limit of very short wavelength (ray propagation). That it indeed "penetrates deeply" is testified to by its fantastic successes in the realms of atomic, molecular, and condensed-state physics. Not the least of these triumphs is the explanation of the periodic table of elements.

The Schrödinger equation determines the wave function ψ (Section 25-6) of a particle, much as the basic equations of Newtonian mechanics determine the particle's classical, precise location. Knowing the wave function means knowing the probability that the particle will be found in the vicinity of any given point of space. Unfortunately, it is mathematically difficult to solve Schrödinger's equation in complete detail; we shall not attempt it even for the hydrogen atom. We can, however, obtain the main features of the wave function in some important cases by applying wave concepts in a somewhat qualitative way. Before tackling the hydrogen atom, which exhibits three-dimensional waves, we are going to look at some simpler classical and quantum-mechanical systems in which the wave motion is confined to one or two dimensions.

26-1 WAVES ALONG A STRING

When one end of a stretched string is shaken, a transverse wave is sent traveling along its length. We have already discussed such one-dimensional waves in Chapter 18, but now we must consider the type of wave motion, called *standing waves,* observed when waves traveling in opposite directions interfere with each other. A string shaken at one end and held fixed at the other end is very responsive to ("resonates with") external disturbances having well-defined, simply related frequencies. This effect, shown in Fig. 26-1, is not difficult to understand. Consider the traveling sinusoidal wave of Fig. 26-2a as it moves to the right along a string or, better, a Slinky; the right end of the Slinky is firmly attached to a massive wall. We know from Section 18-5 that the wave is reflected at the fixed end, with polarity reversed (look back at Fig. 18-13). The upward momentum (crest 1 in Fig. 26-2) at the leading edge of the disturbance is reflected back as downward momentum (trough 1′) after collision with the massive wall. Indeed, one-quarter of a period after the leading edge of the incident wave hits the wall, the leading edge of the reflected wave will have traveled a distance of $\frac{1}{4}$ wavelength

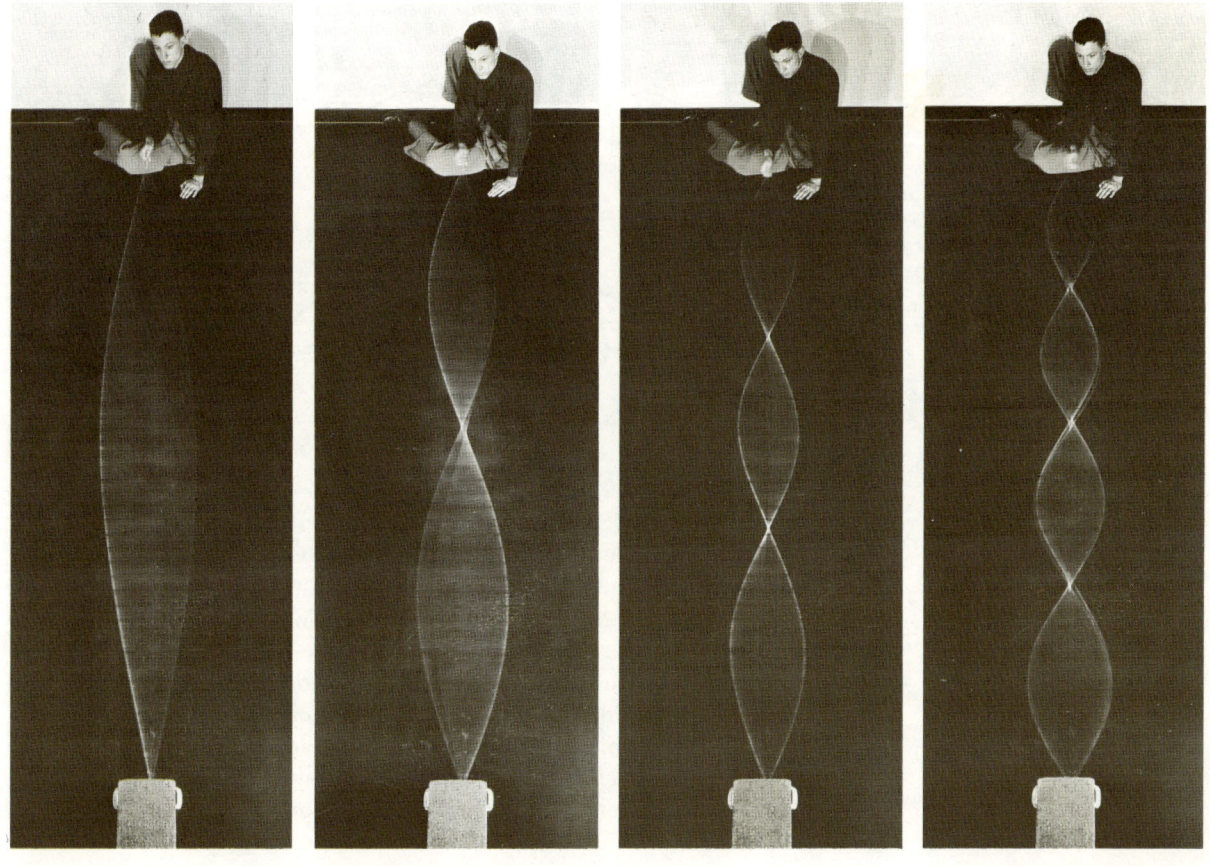

FIGURE 26-1 A resonating string, fixed at one end and vibrated at the other, with standing-wave patterns. (*From PSSC "Physics," D. C. Heath and Company, Boston, 1965.*)

CHAPTER 26 QUANTUM SYSTEMS OF PARTICLES

FIGURE 26-2 A traveling sinusoidal wave along a Slinky being reflected from a massive wall: (a) the wave approaches the wall on the right; (b) one-quarter of a period after the leading edge arrives at the wall, the incident (1) and reflected (1′) crests interfere destructively over $\frac{1}{4}$ wavelength from the wall; (c) a quarter period still later, the incident and reflected waves interfere constructively; (d) still another quarter period later, destructive interference between incident and reflected waves; (e) one more quarter period later, constructive interference.

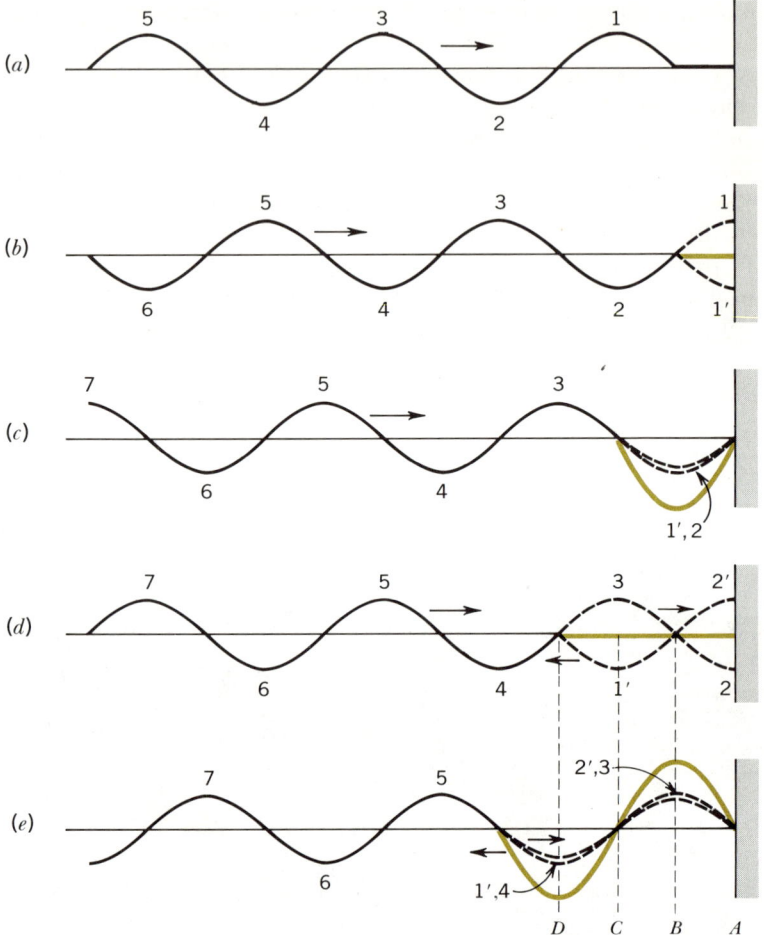

back along the Slinky. At this instant the entire portion of the Slinky within $\frac{1}{4}$ wavelength from the wall will be undisplaced because of the perfect *destructive* interference between the incident and reflected wave (Fig. 26-2b). One-quarter of a period still later trough 1′ of the reflected wave will have moved to the left so that it is perfectly superimposed upon trough 2 of the incident wave moving to the right; as a result there now is complete *constructive* interference between these two troughs, and the net displacement of the Slinky over the first $\frac{1}{2}$ wavelength away from the wall will be double what it would have been for the incident wave alone (Fig. 26-2c). Similarly, still an additional quarter period later, there will again be no displacement near the wall, again because of complete destructive interference between the incident and reflected waves (Fig. 26-2d); and a quarter period after that there will be another large displacement, at least in the vicinity of points B and D, which are $\frac{1}{4}$ and $\frac{3}{4}$ wavelength from the wall. However, there still is no displacement at point A, which is at the wall, and point C, which is $\frac{1}{2}$ wavelength from the wall. Points like A and C, which remain permanently stationary as a result of the interference between the incident and reflected waves, are called *nodes*, while points like B and D, which at certain times undergo maximum disturbance, are called *antinodes*.

Standing waves produced by waves traveling in opposite directions

We now see why nodes exist near the fixed end of the Slinky, but why do we observe a large persistent disturbance of the Slinky only when the nodes divide its length into a whole number of equal pieces (see Fig. 26-1)? This large persistent disturbance occurs when earlier disturbances originally incident upon the fixed end are reflected back to the hand, where they are once more reflected but in such a manner that there is complete constructive interference with the new disturbance just starting out from the hand. The condition for such constructive interference is similar to that for the double-slit experiment in Section 18-7; the *total* distance traveled by the earlier disturbance must be an integral number of wavelengths[1] (or the length of the string or Slinky must be an integral number of half-wavelengths). If this condition does not exist, the interference will not be completely constructive and a final disturbance many times larger than the small disturbance associated with one shake of the hand will not develop. In practice, the net disturbance does not grow indefinitely; rather, an equilibrium situation is reached when the small energy dissipated per unit time by the motion of the string exactly cancels the small energy per unit time being supplied by the hand.

[1] The polarity change during a reflection at the fixed end is exactly compensated for by another polarity change during the reflection from the hand.

But electron waves can also show interference effects. Therefore, consider what the standing-wave patterns shown in Fig. 26-1 mean when they represent the wave function ψ of a particle rather than the transverse displacement of a Slinky (Fig. 26-3a). Recalling that $\psi^2(x)\,\Delta x$ is a measure of the probability of finding the particle in a small interval Δx about the position x, we see that the standing-wave pattern in Fig. 26-1a describes a particle somehow trapped between two walls. The probability of finding the particle adjacent to either wall is zero (it is never there), and although there is some probability that the particle might be at any intermediate position,

Wave-mechanical interpretation of the standing-wave pattern

FIGURE 26-3 (a) The wave function ψ for a particle for the first standing-wave pattern of Fig. 26-1. (b) The probability density ($\propto \psi^2$) corresponding to (a); this photograph, and others like it in this chapter, is of an oscilloscope screen on which the probability density derived from a computer is displayed. (*Robert Ehrlich.*)

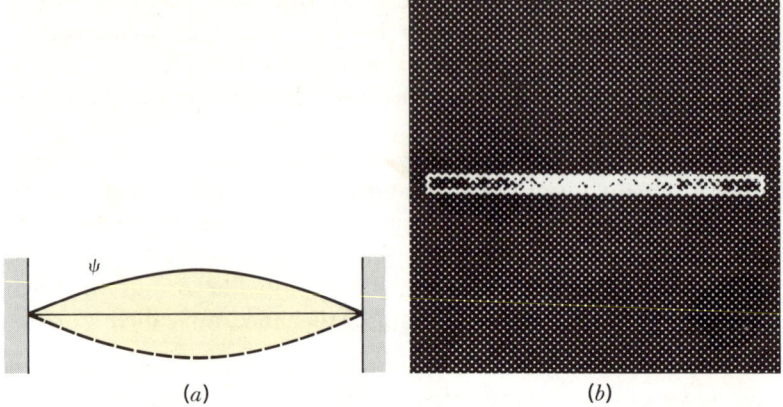

(a)   (b)

Quantization of energy, a consequence of fitting waves at the boundaries

one is most likely to find it very close to the center (where ψ has its antinode). One way to picture this probability (an especially convenient way for the hydrogen atom) is shown in Fig. 26-3b, where the number of points in any interval Δx is proportional to $\psi^2(x)\,\Delta x$, the probability of finding the particle in that interval. Because an integral number of half-waves exist between the ends when particle waves are reflected at the walls, and because a particle's momentum is fixed by its wavelength through the relation $p = h/\lambda$, the existence of the standing wave implies a limitation of allowed momenta and therefore of energies. *Energy quantization appears as a necessary consequence of wave properties.* Unless the particle is like a relatively slow electron (small p, large λ), and unless the distance between the boundaries is as small as the breadth of an atom, the allowed energies are so numerous and so very closely spaced as to be effectively continuous.

26-2 WAVES OVER A CIRCULAR MEMBRANE

Standing waves on a kettledrum membrane

We turn to another macroscopic system, a stretched sheet of rubber rigidly fixed at its circular boundary. Two-dimensional waves are propagated over such a membrane. They are reflected when they meet the boundary, and once again only those wave patterns can persist which correspond to standing waves. A photograph of an observed pattern is shown in Fig. 26-4 along with a sketch of this and some additional patterns.

The simplest kind of standing waves on the membrane are shown in Fig. 26-5. Here we may think of circular waves traveling radially outward and inward (rings expanding and contracting) and inter-

SECTION 26-2
WAVES OVER
A CIRCULAR
MEMBRANE

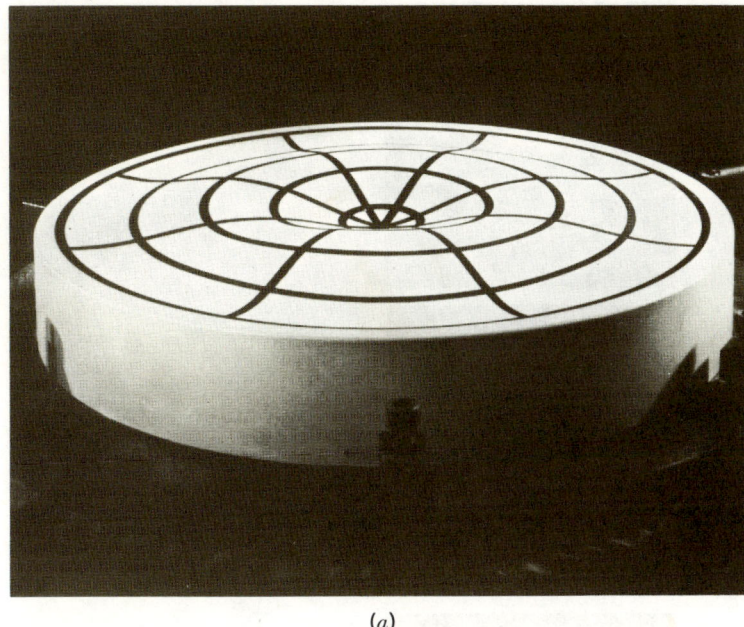

FIGURE 26-4 Standing waves on a circular membrane: (a) A pattern on a drumhead induced by a loudspeaker beneath ($j = 2$, $k = 0$); (*Harvard Project Physics and National Film Board of Canada.*) (b) top view of a membrane for another standing-wave pattern, the plus and minus signs indicating, respectively, portions above and below the equilibrium plane ($j = 1$, $k = 1$); (c) a number of standing-wave patterns identified by j (number of nodal circles) and k (number of diametrical nodal lines).

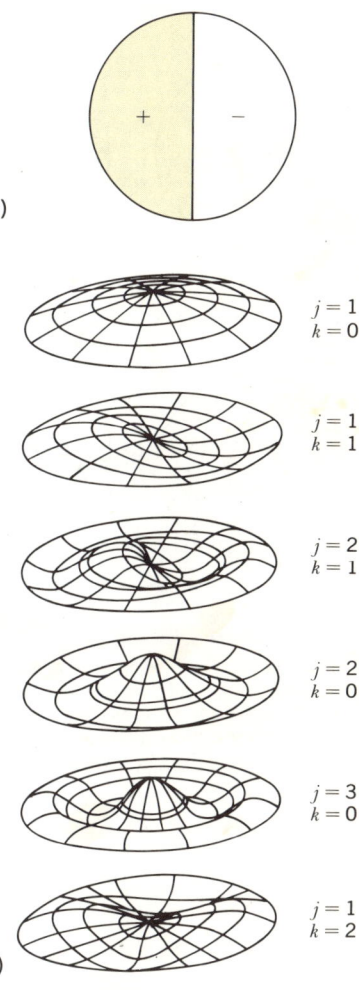

fering to form standing patterns. The outer edge of the membrane is a nodal circle, i.e., each of its points is a node, but the center, free to oscillate, is an antinode. The resonant patterns can be labeled by the integer $j = 1, 2, 3, \ldots$, which is the "quantum" number giving the number of nodal circles in the progressively more complicated patterns. (An integral number of half-wave patterns again fits between the center of the membrane and the outer edge, but detailed calculations show that the resonant frequencies of the several oscillation modes are not exactly in the ratio of integers, as is the case for the standing waves on a Slinky.)

In Fig. 26-5a the plus and minus signs indicate, from a top view of the membrane, the portions which are above or beneath the equilibrium plane at one instant; half a cycle later, the signs are interchanged. The wave functions for the symmetrical modes are shown in Fig. 26-5b; here the displacement $R(r)$ of points of the membrane from the equilibrium plane is plotted as a function of the distance r from the center. Each solid line represents $R(r)$ at an instant of time, and the shading indicates the range of possible excursions.

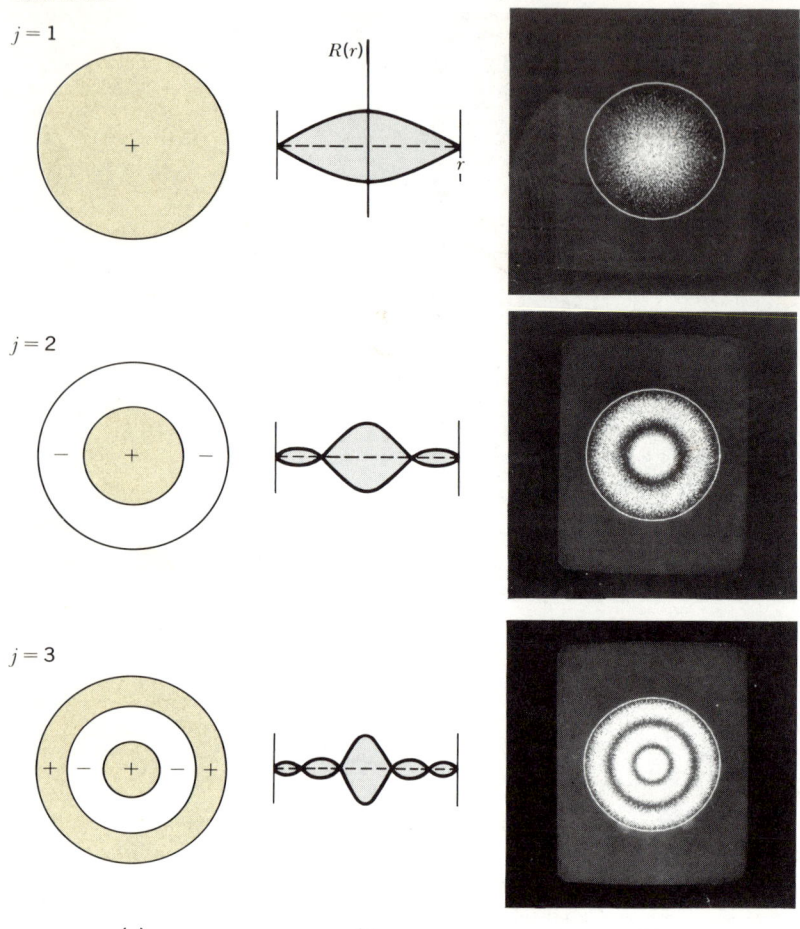

FIGURE 26-5 (a) Standing-wave patterns, (b) the radial factors of the wave function ψ, and (c) probability densities $[\propto \psi^2 \propto R^2(r)]$ for the first three symmetrical modes. *(Robert Ehrlich.)*

How would we interpret the standing-wave patterns for the kettledrum if they represented the wave function ψ of a particle rather than the transverse displacement of the membrane? Wherever ψ is zero, the probability of finding the particle is zero; where ψ is large, the probability is high. Thus, the particle is confined to circular rings concentric to the outer boundary, as shown in Fig. 26-5c.

We now move to the standing-wave patterns for the membrane (or particle in a barrel) which exhibit an angular dependence. Some of these asymmetrical modes, i.e., modes having angular dependence,

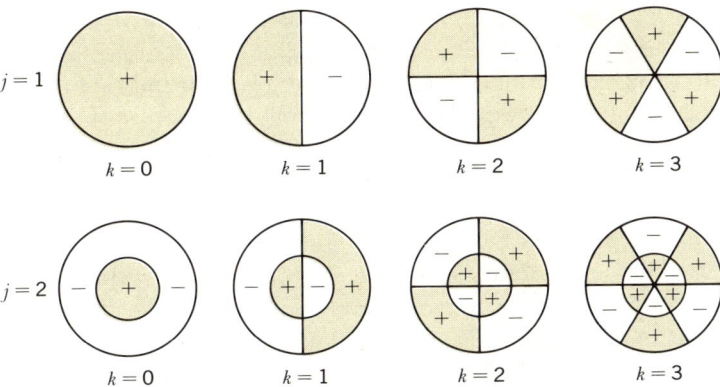

FIGURE 26-6 Standing-wave patterns for a circular membrane; j gives the number of nodal circles, and k gives the number of diametrical nodal lines.

are shown in Fig. 26-6; they are characterized by diametrical nodal lines, as well as circular nodal lines. The angular dependence of standing-wave patterns is associated with an additional integral quantum number, $k = 0, 1, 2, \ldots$, the value of k being the number of diametrical nodal lines.

More details of the pattern corresponding to $j = 1$ and $k = 1$ are shown in Fig. 26-7. The displacement of the membrane is proportional to the radial factor $R(r)$ shown in Fig. 26-7a. Because the wave pattern is not symmetrical around the center of the membrane, the proportionality factor for R will be different for different diametrical lines. In other words, it will depend on the *azimuthal angle* ϕ defined in Fig. 26-7b. The angle ϕ is measured in the equilibrium plane of the membrane relative to some arbitrary axis x. Denoting the azimuthal factor Φ, we then have $R(r)\Phi(\phi)$ for the displacement of the membrane as it varies with the orientation of the diametrical line (given by ϕ) and with the distance out from the center (given by r). Figure 26-7c shows how the azimuthal factor Φ varies with ϕ.

If the standing-wave pattern in Fig. 26-7 represented a wave function ψ of a particle, then because Φ is zero at $\phi = \pm 90°$, the wave function ψ vanishes along a single nodal diameter which is at right angles to the x axis. A plot of the probability density $\psi^2 \propto R^2(r)\Phi^2(\phi)$ is shown in Fig. 26-8. As before, the regions with the greatest density of points specify locations for which the probability of locating the particle is a maximum.

Like the standing waves along a string (Fig. 26-1), the standing wave of Fig. 26-7b can be thought of as the superposition of two waves traveling in opposite directions. In fact, imagine that the pattern of

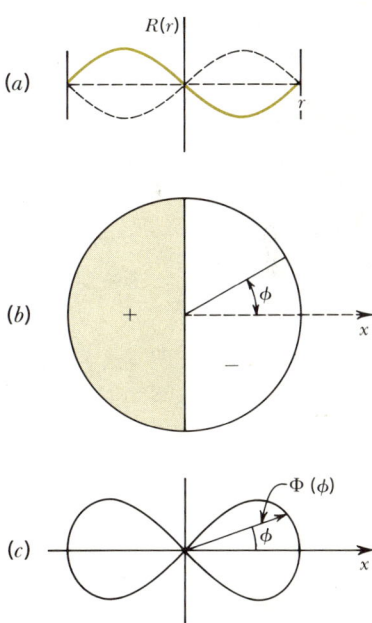

FIGURE 26-7 Details of the standing-wave pattern for the asymmetrical $j = 1$, $k = 1$ mode: (*a*) side view of the radial factor $R(r)$; (*b*) top view, with the azimuthal angle ϕ defined; (*c*) the azimuthal factor $\Phi(\phi)$.

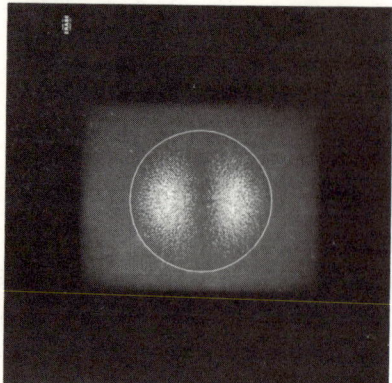

FIGURE 26-8 Probability density $[\propto \psi^2 \propto R^2(r)\Phi^2(\phi)]$ for the $j = 1$, $k = 1$ mode shown in Fig. 26-7. *(Robert Ehrlich.)*

Fig. 26-1b is wrapped around a circle and the two ends joined together. The result is exactly the standing-wave pattern of Fig. 26-7b at a constant distance from the center of the membrane. Therefore, the standing-wave pattern on the circular membrane is just what we would get from the interference of two circular waves sweeping around the center of the membrane in opposite senses. Indeed, we can think of the angular patterns for the two circulating waves as rotating in the fashion shown in Fig. 26-9, one in the clockwise and the other in the counterclockwise sense. Moreover, if these circulating waves represent the wave function ψ of a particle, the probability density for the particle (Fig. 26-8) can also be considered the standing-wave pattern of two rotating patterns. In this way the angular dependence of the asymmetrical standing-wave pattern is associated with the angular motion (momentum) shown in Fig. 26-10.

26-3 THE PUDDLE AND THE PARTICLE IN A SAUCER

Before proceeding to the wave-mechanical description of hydrogen, we need to look at an example somewhat more complicated than the drumhead. We consider (1) waves traveling along the surface of a liquid in a saucer or the wave-mechanical analog (2), a particle sliding in a saucer (see Fig. 26-11). You may have noticed at the

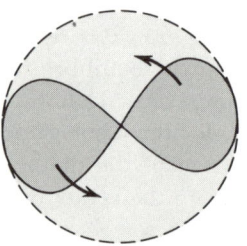 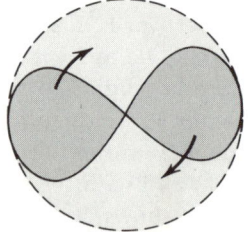

$j = 1$, $k = 1$

FIGURE 26-9 Two circulating waves, one clockwise and one counterclockwise, giving rise to the asymmetrical probability density for the $j = 1$, $k = 1$ mode of Fig. 26-8.

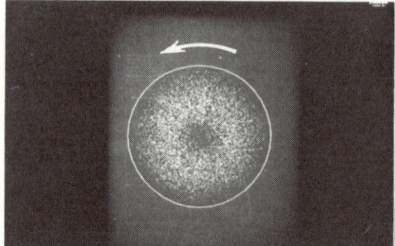 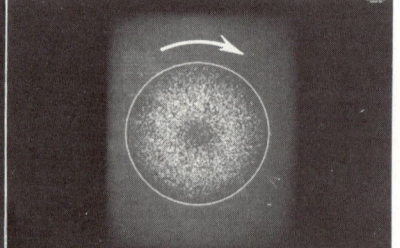

FIGURE 26-10 Probability densities for angular motion (angular momentum) in the clockwise and counterclockwise senses; their combination yields the asymmetrical pattern of Fig. 26-8. *(Robert Ehrlich.)*

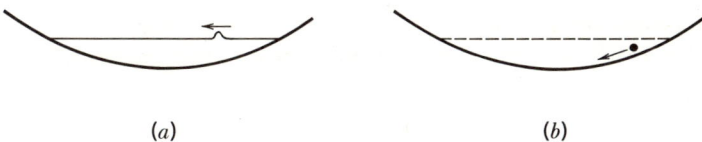

(a)   (b)

FIGURE 26-11 (a) Waves traveling along the surface of a liquid in a saucer and (b) the wave-mechanical analog, a particle sliding in a saucer.

beach that the water waves change in wavelength as they approach the shoreline. There is a similar behavior for a wave mechanical particle confined to a saucer: as the particle goes uphill and loses speed, its wavelength $\lambda = h/mv$ increases.

Apart from the marked variation in wavelength with distance from the center of the saucer, the allowed wave patterns here are much like those for the circular membrane (see Fig. 26-12). Again there are symmetrical wave functions, with no angular dependence, and asymmetrical waves with diametrical nodal lines. We may associate the symmetrical modes with water waves which travel in the radial direction (with circular wavefronts). In the wave-mechanical analog the symmetrical solutions correspond to situations in which the particle is sliding along radial lines so as to pass through the center of the saucer; it therefore has zero angular momentum (relative to the center point). On the other hand, the asymmetrical solutions correspond to a particle coasting in a loop; here the particle's angular

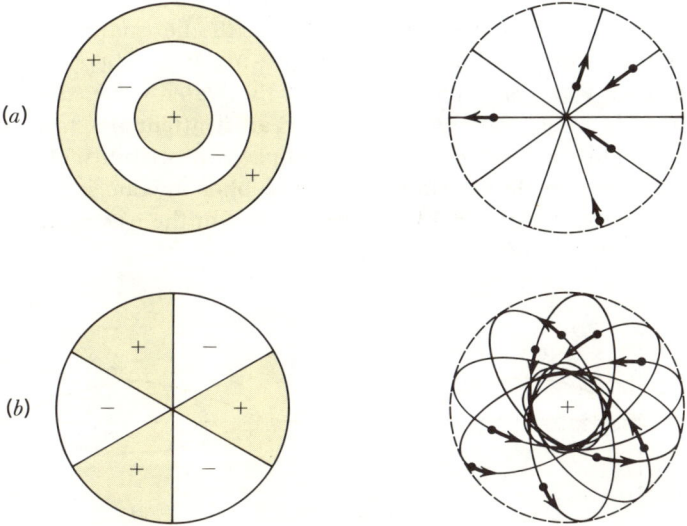

FIGURE 26-12 (a) Standing-wave patterns and (b) the corresponding particle trajectories for symmetrical modes (particle with zero angular momentum) and asymmetrical modes (particle with nonzero angular momentum).

momentum is not zero. Thus again there is a physical distinction between symmetrical and asymmetrical wave patterns.

For the symmetrical modes in Fig. 26-12, with the particle conceived to move through the center, we see not merely one path but a whole collection. This is to indicate that for a given total energy of the particle in the saucer any one of these paths is equivalent to any other. For a *classical* particle in a saucer the conservation of energy imposes an upper limit on the particle's distance from the center. Each path has two turning points where the particle comes to rest, so that the limits of all paths of a fixed energy define a circle. For the asymmetrical modes the particle can be thought to have been projected initially in such a direction that it missed passing through the saucer center; thereafter it executes a loop enclosing the center. For a given total energy and angular momentum the collection of all possible loops now defines two circles, the outer one corresponding to the farthest point the particle travels from the center, and the inner one to the closest approach to the center. This "hole in the middle," a consequence of angular-momentum conservation for those situations in which the angular momentum differs from zero, has its counterpart in the allowed patterns for water waves in a saucer or waves along a membrane: all asymmetrical modes have a nodal point at the center, in contrast to the antinode for symmetrical modes (see Fig. 26-12).

Query How do we know that the particle, having once missed the center, always misses the center?

We have shown that for a *classical* particle in a saucer, and the same holds true for a puddle, there exists an impassable barrier, located by the points at which, for a given total energy, the classical kinetic energy vanishes. Here the analogy with the quantum-mechanical problem breaks down in a strange way. It can be shown that the matter wave of a particle, unlike the water wave, does not drop exactly to zero at the barrier; instead it attenuates to zero farther out. This means that there is a nonzero probability of finding the particle outside the classical limit, a phenomenon known as *barrier penetration*. Figure 26-13 shows the effect in the wave pattern for the first symmetrical mode.

Barrier-penetration phenomenon

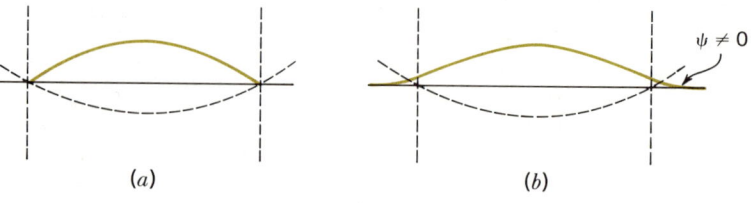

FIGURE 26-13 Allowed wave patterns for (*a*) waves along liquid surface in saucer and (*b*) wave-mechanical particle in a saucer. The cross section of the saucer is shown with a dashed line.

Now we come to the wave-mechanical problem of an electron attracted to a proton by the electric force varying inversely as the square of the distance between them. The corresponding electrostatic potential energy E_p is shown in Fig. 26-14 as a function of the electron's distance r from the central proton. For a given constant total energy, $E = E_p + E_k$, the electron kinetic energy E_k drops to zero at points A and A'. Thus, the three-dimensional energy barrier is a spherical surface of diameter AA', and, classically, the electron is trapped in this sphere. Since the electron's kinetic energy and speed decrease as its distance from the center increases, the electron wavelength is long at large distances from the nucleus and short close to it.

26-4 THE WAVE THEORY OF THE HYDROGEN ATOM

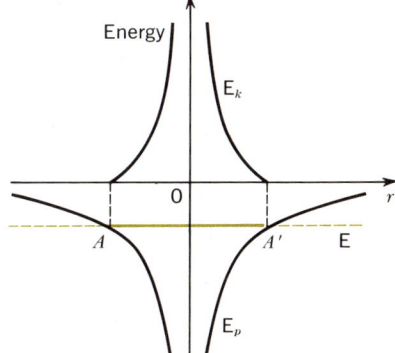

FIGURE 26-14 The potential energy E_p, the kinetic energy E_k, and the constant total energy E for an electron attracted by a proton plotted as a function of the electron's distance r from the proton. A and A' are points where the electron's kinetic energy is zero.

We can anticipate that the features we have already met with in simpler one- and two-dimensional problems will reappear here: waves reflected from boundaries (sharply defined *or* continuous) producing standing waves; quantum numbers, one for each dimension, giving the number of half-waves that can be fitted within the boundaries; symmetrical wave patterns corresponding to zero angular momentum and asymmetrical wave patterns with a net angular momentum; and quantization of the system's energy as a consequence of the limited number of ways in which waves can be fitted within the boundaries of the system. Indeed, precisely these features emerge once the Schrödinger wave equation is solved.

The first three symmetrical wave patterns for the hydrogen atom are shown in Fig. 26-15. Here we see the shape of the standing wave for ψ along any fixed radius from the center of the atom. The three corresponding electron density patterns are shown in cross section in Fig. 26-16. A single quantum number $n = 1, 2, 3, \ldots$ identifies these symmetrical solutions for ψ; n counts the number of zeros (nodes) in ψ along a radius or, equivalently, the number of shells in the probability-density pattern. We notice that the wavelength, as indicated by the separation of adjacent zeros in ψ, decreases as we go from $n = 1$ to $n = 3$. This suggests that the atom's total energy increases with n. Detailed computations from the Schrödinger equation show, in fact, that the atom's energy is given by $E = (-13.6 \text{ eV})/n^2$, precisely the relation [24-10] emerging from the Bohr atomic theory. The quantum number n is called the *principal quantum number*; it specifies the atom's energy.

Principal quantum number n specifies hydrogen atom's energy

For $n = 1$, the atom is in its ground state. We may think of the electron as forming a fuzzy ball centered at the nucleus. For each excited state, an additional fuzzy shell of electron probability ap-

604

CHAPTER 26
QUANTUM
SYSTEMS OF
PARTICLES

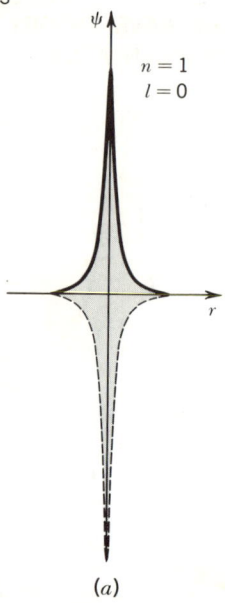
(a)

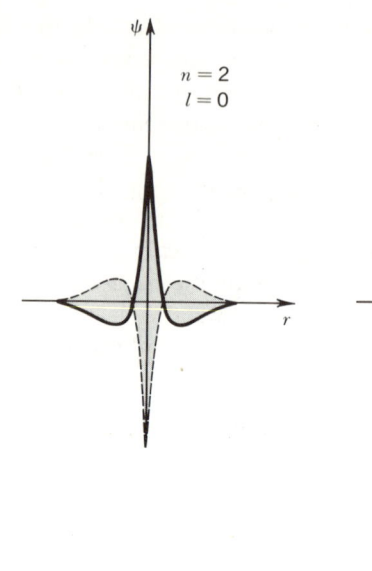
(b)

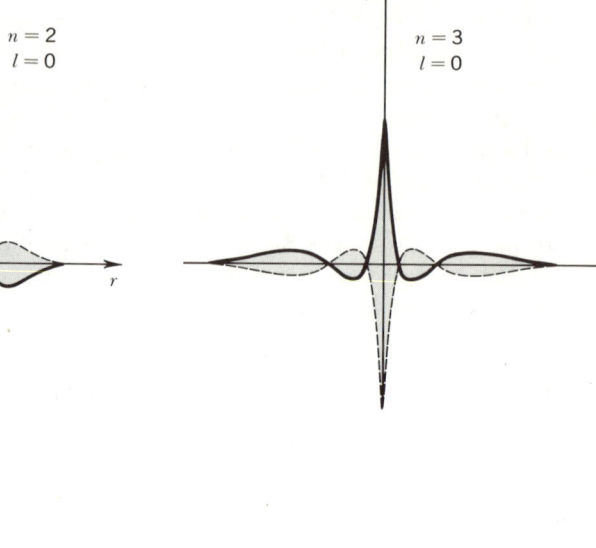
(c)

FIGURE 26-15 The first three symmetrical wave functions ψ for the hydrogen atom plotted as a function of the electron's distance r from the proton.

FIGURE 26-16 The probability densities for the first three ($n = 1, 2, 3$) symmetrical wave functions shown in Fig. 26-15. Each is, in effect, a slice through the center of the atom. (*Robert Ehrlich.*)

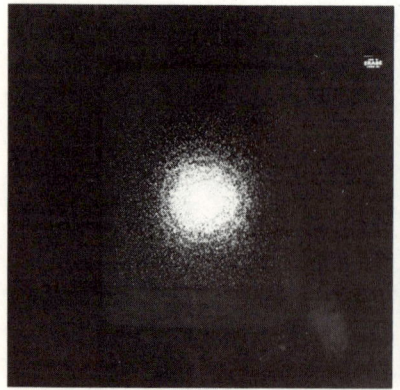
(a)

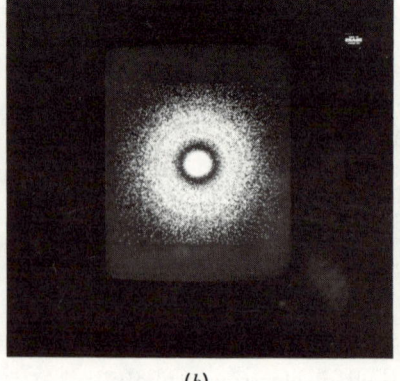
(b)

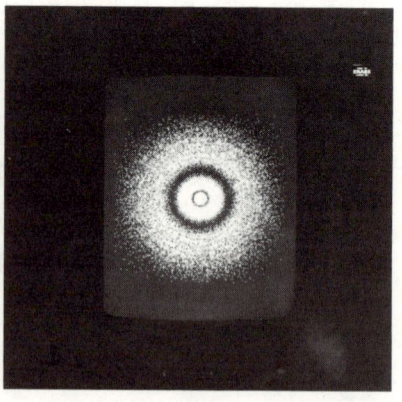
(c)

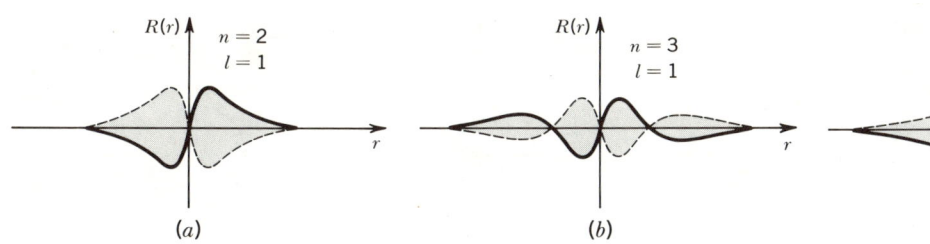
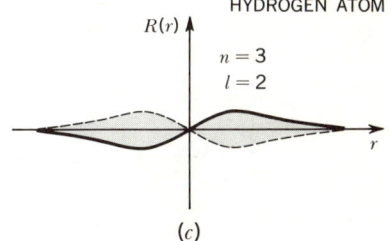

FIGURE 26-17 The variation with distance from the nucleus of asymmetrical wave functions for the hydrogen atom.

pears. Owing to their fuzziness, the shells cannot be assigned precise radii; in other words, there are no such things as definite Bohr orbits for the electron. Nevertheless, the rough dimensions of the shells, of the order of 1 Å, are in harmony with the Bohr model.

Let us now examine the asymmetrical wave patterns. Figure 26-17 shows, for $n = 2$ and $n = 3$, how the wave function ψ varies as we go outward from the nucleus along any radius. It is seen that the form of the radial dependence of ψ now depends on two quantum numbers, the principal quantum number n and an additional number l, which can take on the values $0, 1, 2, \ldots, n - 1$. (Solutions with $l = 0$ are all symmetrical and therefore do not appear in Fig. 26-17. In particular, there is no asymmetrical solution $l \neq 0$ for $n = 1$.) In all asymmetrical cases $\psi = 0$ at $r = 0$. Because angular-momentum conservation implies that an electron with initial angular momentum can never be found at the nucleus, we can associate, just as with a particle in a saucer, presence at the center with zero angular momentum and complete evasion ($\psi = 0$) of the center with finite angular momentum. Thus, we associate spherical symmetry ($l = 0$) with zero angular momentum and asymmetry, or angular dependence ($l \neq 0$), with finite angular momentum.

For a complete picture of ψ, we also need to know how it varies when r is held fixed but the angular position with respect to the nucleus changes. The proportionality factor $Y(\theta,\phi)$, which gives the angular variation, is shown in the three-dimensional polar graphs of Fig. 26-18. Since it takes two coordinates (latitude θ, longitude ϕ) to fix angular position in three dimensions, it is not surprising that in the solutions of the Schrödinger equation for the hydrogen atom the angular part of ψ depends on two quantum numbers. One of these is l, which entered in the radial dependence; the other is the quantum number m_l, which for each l takes on the integral values $0, \pm 1, \pm 2, \ldots, \pm l$. The linking via l of the radial and angular dependence of ψ is a new feature, not encountered in our examination of two-dimensional waves. Because $l = 0$ characterizes a spherically sym-

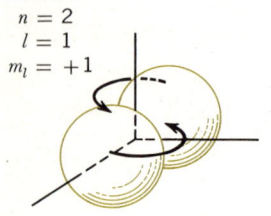
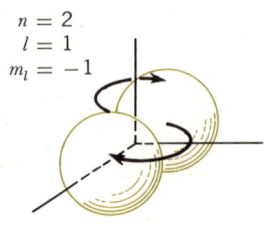
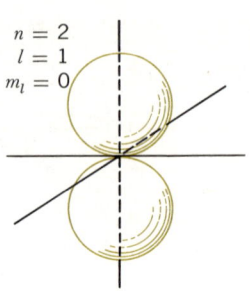
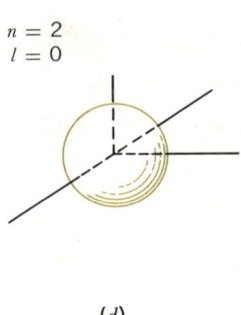

(a)　　　　　　(b)　　　　　　(c)　　　　　　(d)

FIGURE 26-18 Angular portions of the wave functions for the (a) $m_l = +1$, (b) $m_l = -1$, (c) $m_l = 0$, all for $l = 1$. (d) The angular portion of the wave function for $l = 0$.

Orbital quantum number l specifies orbital angular momentum.

Quantum number m_l specifies orientation of electron "orbit."

metrical wave function, there is no angular dependence and Fig. 26-18d is simply a sphere.

The principal quantum number n governs the energy of the atom. All the l states corresponding to the same value of n have the same energy. We call l the *orbital quantum number* because it relates to the angular momentum of the atom, which arises, classically speaking, from the orbiting of the electron around the nucleus. In fact, wave mechanics shows that the magnitude of the orbital angular momentum is given by $\sqrt{l(l+1)}\,\hbar$. Here we use the symbol $\hbar$ (read as "h-bar") to denote $h/2\pi$.

The quantum number m_l might be said to give the orientation of the plane in which the electron orbits. For the states $m_l = +1$ and $m_l = -1$ the angular part of the wave function can be thought of as a pair of spheres revolving about the z axis and sweeping out a doughnut-shaped surface. In Fig. 26-18a we see the pair of spheres revolving in one sense (the $m_l = +1$ state), while in Fig. 26-18b they revolve in the opposite sense (the $m_l = -1$ state). The behavior here is like that shown before (Fig. 26-9) for standing waves on a circular membrane. If the wave patterns rotate, then so do the corresponding electron densities, and we can interpret the $m_l = +1$ state as corresponding to an electron which orbits the nucleus in a plane perpendicular to the z axis in one sense, while in the $m_l = -1$ state the electron orbits in the opposite sense.

A more difficult question is the interpretation of the angular pattern shown in Fig. 26-18c for the $m_l = 0$ state. It consists of a pair of nonrotating spheres centered on the z axis. For this state the electron still has the same magnitude of angular momentum as for the two revolving spheres shown in Fig. 26-18a and b. The pattern in Fig. 26-18c is a standing-wave pattern, one produced by the interference of traveling waves. Indeed, it arises from the interference of a number

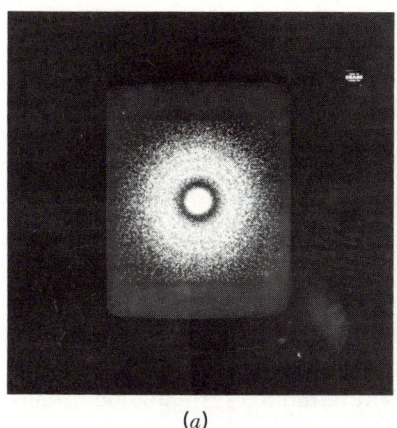 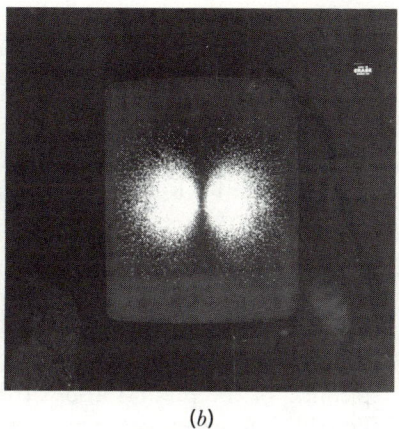 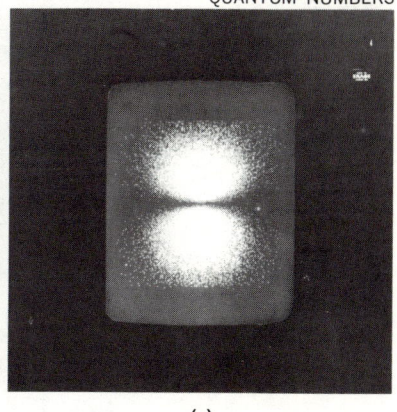

(a)   (b)   (c)

FIGURE 26-19 Probability densities as slices through the center of the hydrogen atom for (a) $n = 2$, $l = 0$, (b) $n = 2$, $l = 1$, $m_l = \pm 1$, and (c) $n = 2$, $l = 1$, $m_l = 0$. (*Robert Ehrlich*.)

of revolving spheres, like those of Fig. 26-18a and b, but now moving around axes lying in the xy plane. These axes may take on all possible orientations in the xy plane; moreover, for any given axis we again have both clockwise and counterclockwise revolution of the sphere. When we add together all the resulting patterns, we are left with the two spheres shown in Fig. 26-18c. The waves in the equatorial plane have been washed out by destructive interference. The standing-wave pattern implies, then, that the electron has angular momentum (just as much as for the $m_l = \pm 1$ states of Fig. 26-18a and b) about an axis somewhere in the xy plane. However, the orientation of the axis within this plane is left completely undetermined.

As before, the wave patterns grow more complex as the value of the principal quantum number increases. The electron density patterns for some of these states are shown in Fig. 26-19. Note that if we add together the density patterns, e.g., for all three possible values of m_l (0 and ± 1) for $l = 1$, we get a spherically symmetric charge distribution: the two knobs along the z axis fit exactly into the two indentations of the doughnut. This illustrates the general rule: the component m_l states for nonzero angular momentum are individually angular-dependent, but these separate density patterns add to give a spherically symmetric pattern or "shell," corresponding to zero angular momentum.

26-5 ELECTRON SPIN AND THE FOUR QUANTUM NUMBERS

When standing waves are fitted in one dimension, a single quantum number emerges to characterize the restrictions imposed by the boundaries. For two-dimensional waves, as on the circular mem-

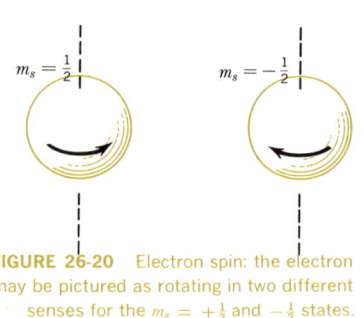

FIGURE 26-20 Electron spin: the electron may be pictured as rotating in two different senses for the $m_s = +\tfrac{1}{2}$ and $-\tfrac{1}{2}$ states.

Spin orientation quantum number m_s

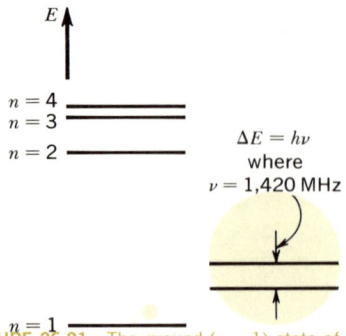

FIGURE 26-21 The ground ($n = 1$) state of hydrogen actually consists of two slightly separated energy levels whose difference corresponds to a photon of 1,420 MHz.

brane, two quantum numbers are needed to identify the possible wave patterns; and, as we have just seen for the hydrogen atom, standing electron waves extending into three dimensions about the nucleus require three quantum numbers, n, l, and m_l. Though we might suppose that this exhausts the possible quantum numbers, there is actually a fourth, associated with *electron spin.*

A variety of experiments show that an electron always has angular momentum in addition to that arising (in classical language) from its orbiting the nucleus. One can draw a comparison with the solar system, which has, over and above the angular momentum arising from the planets orbiting the sun, angular momentum from the spinning of the planets about their internal axes. By analogy, then, the additional electron angular momentum, independent of the state of motion of the electron relative to other objects, is called *spin* angular momentum or simply *spin*. If we insist upon having a classical picture of electron spin, we can imagine the electron as a little ball of electric charge spinning about an internal axis. The orbital angular momentum for $m_l = +1$ or -1 states can be thought of as arising from the clockwise or counterclockwise orbital motion of the electron in the equatorial plane. Similarly, in the case of electron spin we may picture the electron as spinning in either of two senses (Fig. 26-20). Unlike orbital angular momentum with quantum number 1 and three orientations ($+1, 0, -1$), electron spin has only two orientations; but these orientational quantum numbers still differ by the integral number 1. Moreover, the spin angular momentum of the electron is found to be $\tfrac{1}{2}\hbar$. We therefore introduce a *spin-orientation quantum number* m_s with only two possible values, $m_s = \tfrac{1}{2}$ or $m_s = -\tfrac{1}{2}$.

A proton, too, is found to have a spin of magnitude $\tfrac{1}{2}\hbar$, again with two rotation senses. Now, a spinning positive or negative charge can be thought of as a little loop of electric current, i.e., a little magnet. Thus, apart from the energy levels in a hydrogen atom arising from the orbiting of the electron, as a wave, about the proton, we have the possibility for very slight energy differences originating from differences in relative orientation of the proton and electron, regarded as tiny magnets. For a given orientation of the proton there are two possible orientations of the electron, corresponding to $m_s = \tfrac{1}{2}$ and $-\tfrac{1}{2}$. The energies of these states should differ slightly. In fact, for the ground state ($n = 1$) of hydrogen the energy difference $\Delta E = h\nu$ corresponds to a photon having a frequency of about $1{,}420 \times 10^6$ Hz or a wavelength of about 21 cm (see Fig. 26-21). Such radiation lies in the long-wavelength radio portion of the electromagnetic spectrum, in contrast to the visible-light photons from changes in n. Indeed, a prime source of information in radio astronomy is just the 21-cm line emitted by intergalactic clouds of hydrogen, which may be invisible to optical telescopes.

The effects of electron spin are evident not only in the splitting of the $n = 1$ level of hydrogen but also in the appearance of *fine structure* (closely spaced energy levels) for still higher states. For energy levels with $n > 1$ the multiplicity of levels over that given by the Bohr theory must also be attributed to a subtle interaction between electron spin and orbital angular momentum.

To recapitulate, four quantum numbers specify the state of the hydrogen atom: (1) the principal quantum number n, which gives the gross energy of the atom; (2) the orbital quantum number l, which determines the magnitude of the electron's orbital angular momentum; (3) the orbital-orientation quantum number m_l, which has to do with the orientation of the election's orbital motion; and (4) the spin-orientation quantum number m_s, which labels the two possible orientations of the electron's spin. Any attempt to describe the quantum numbers m_l and m_s classically is of necessity not very successful; the electron does not really move in a fixed plane, nor is it in fact a spinning ball of charge. However, this is the best we can do in classical language.

To specify an atomic state, give the quantum numbers.

An energy-level diagram for hydrogen, similar to that in Fig. 24-5, is given in Fig. 26-22. The figure shows not only the allowed atomic energies but also counts the number of quantum states corresponding

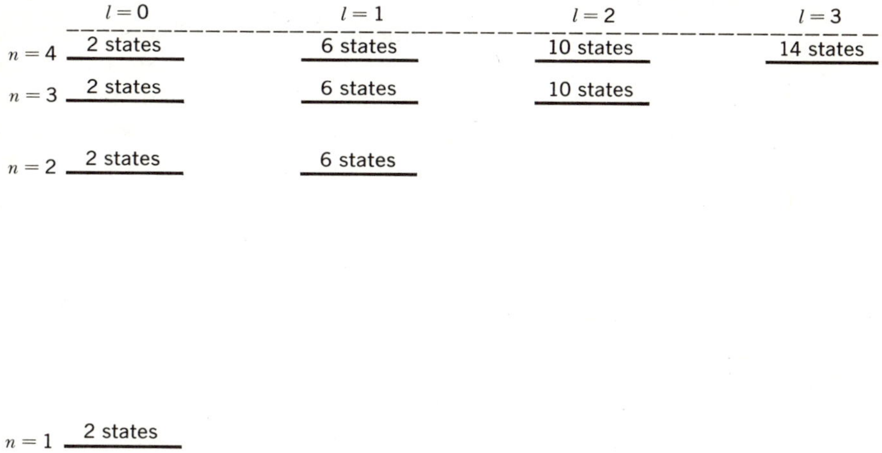

FIGURE 26-22 Energy-level diagram for hydrogen with quantum states identified according to n and l. Each level consists of the number of states indicated.

	$l = 0$	$l = 1$	$l = 2$	$l = 3$
$n = 4$	2 states	6 states	10 states	14 states
$n = 3$	2 states	6 states	10 states	
$n = 2$	2 states	6 states		
$n = 1$	2 states			

to each energy level. The count can be obtained as follows. Start with the ground state, $n = 1$. The only possible value for l is 0; m_l must then also be 0. There remain the two possible values, $\frac{1}{2}$ and $-\frac{1}{2}$, for m_s. All told, then, there are two states for $n = 1$. Moving to the first excited state, $n = 2$, we again have two states for $l = 0$. Now we may also have $l = 1$, with $m_l = -1, 0$, or $+1$; and for each value of m_l there are the two possible values, $\frac{1}{2}$ and $-\frac{1}{2}$, for m_s. Therefore, with $l = 1$, we have six states altogether. Moving up to $n = 3$, we again have two states for $l = 0$ and six states for $l = 1$. In addition, we now have $l = 2$ and $m_l = -2, -1, 0, 1$, or 2; and for each of these five, the two possible values for m_s, giving ten $l = 2$ states in all. Still higher entries in the figure are completed in similar fashion.

A hydrogen atom may exist in any of the states shown in Fig. 26-22. The atom has its lowest energy in the ground ($n = 1$) state. If the atom is initially in a higher energy state, it may make downward transitions, radiating the characteristic hydrogen spectrum.

26-6 THE PERIODIC TABLE

Now we want to explore the structure of atoms having more than one electron, in particular to see how the existence of the four quantum numbers, together with the operation of an additional fundamental principle, leads to an understanding of the chemical properties of the elements as displayed in the periodic table.

A somewhat more complex atom than hydrogen is lithium, $_3$Li, with a triple positive charge at its nucleus and three electrons when electrically neutral. The spectrum radiated by lithium, among other evidence, establishes that when the atom is in its most stable configuration only *two* of the electrons are in the state $n = 1$; the third is in the $n = 2$ state. Offhand, one would imagine that the atom would be in its ground state only when all three electrons were individually in their ground states; i.e., the three electrons should all carry the quantum numbers $n = 1$, $l = 0$, $m_l = 0$, with only the m_s values possibly differing among them.

The actual state of affairs illustrates a fundamental principle proposed by W. Pauli (1900–1958) in 1925. The *Pauli exclusion principle* asserts that *no two electrons in a single atom occupy the same state* or, equivalently, *no two electrons of the same atom have the same set of four quantum numbers*. It can be thought of, so to speak, as legislation against overcrowding. Thus for $_3$Li the $n = 1$ state is completed with:

> The Pauli exclusion principle: no two electrons in the same atom have the same set of four quantum numbers.

Electron 1: $n = 1$ $l = 0$ $m_l = 0$ $m_s = -\frac{1}{2}$
Electron 2: $n = 1$ $l = 0$ $m_l = 0$ $m_s = \frac{1}{2}$

The lowest energy level available to the third electron in $_3$Li is then that for which $n = 2$. For an atom of $_2$He, with only two electrons when electrically neutral, we then have the $n = 1, l = 0$ *shell* completely occupied.

The Pauli exclusion principle, allowing only one electron in any one of the states shown in Fig. 26-22, means that as the atomic number (nuclear charge) of the chemical element increases and the number of electrons increases correspondingly, the additional electrons are forced to occupy increasingly higher states. On this basis, we can arrange the elements according to the number of available states shown in Fig. 26-22, to arrive at Fig. 26-23. In Fig. 26-23 the position of the element symbol gives the quantum numbers of the "highest" electron when the atom is in its ground state. The periodicity in the chemical properties is related to the periodicity in the filling of electron shells. For example, as we go from $_5$B to $_{10}$Ne, the electrons fill, one by one, the six states available for $n = 2$ and $l = 1$. That these six electrons actually form a closed shell, or spherically symmetrical charge distribution, is evident upon reexamining Fig. 26-18, which shows electron densities for the $m_l = 0$ and $m_l = \pm 1$ states. Adding all these together, we see qualitatively (and this is confirmed by wave-mechanical calculations) that the space around the nucleus is symmetrically filled with electrons (see Fig. 26-24). Thus neon, with a total of 10 electrons, should be (and is) similar to helium, i.e.,

Similar electron configuration, similar chemical properties

FIGURE 26-23 The chemical elements ordered according to the quantum numbers associated with the "highest" electron when the atom is in its ground state; in effect, the periodic table.

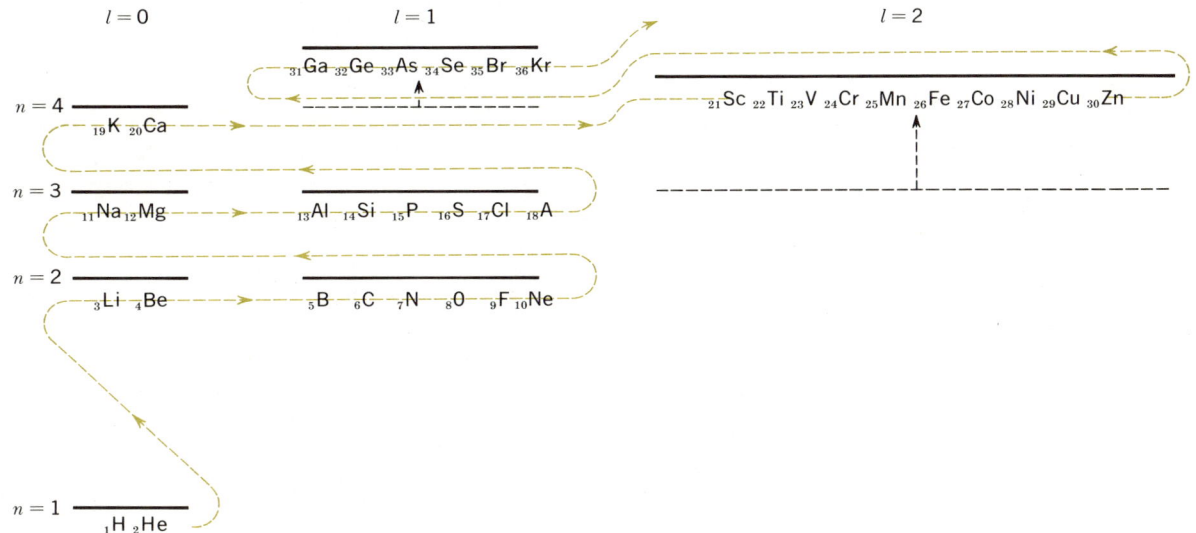

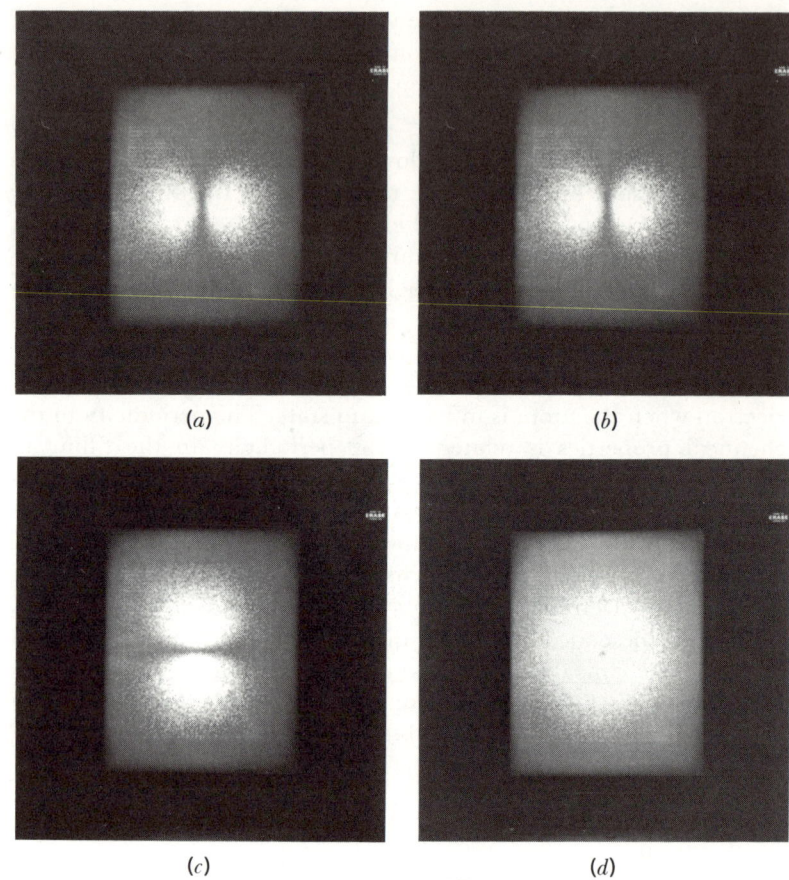

FIGURE 26-24 The combination of the probability densities for the (a), (b) $m_l = \pm 1$ and (c) $m_l = 0$ states yields (d) a spherically symmetrical charge distribution. (*Robert Ehrlich.*)

chemically inactive. More generally, all elements in a single column have similar electron configurations and therefore similar chemical properties.

The step-by-step filling of the n shells does not proceed with perfect regularity. The observed chemical properties of $_{19}$K show that the 19th electron goes into an $n = 4$, $l = 0$ state rather than an $n = 3$, $l = 2$ state. This means that the 19th electron actually has less energy in an $n = 4$, $l = 0$ state than in an $n = 3$, $l = 2$ state. Only after $_{20}$Ca completes the first subshell of $n = 4$ is the $n = 3$ shell filled out with "inside" electrons of the *transition elements*, atomic numbers 21 to 30. A similar shift occurs with elements 31 to 36.

26-7 GROUPS OF ATOMS

We have now seen, at least qualitatively, how the properties of atoms, particularly their stability, can be accounted for solely by quantum concepts based on the wave properties of particles. Likewise for *groups* of atoms. It is true that the forces between atoms are primarily electrical: individual molecules, as well as all states of matter, consist of large numbers of particles, negatively and positively charged, attracting and repelling each other. Yet classical electromagnetic theory by itself is quite insufficient to deal with the incredibly diverse behavior of such assemblages of atoms. Any basic explanation must invoke the wave properties of the fundamental particles.

The physics (or chemistry or molecular biology or radio astronomy) of molecules and the condensed states of matter extends far beyond the scope of this book. Providing a glimpse of how wave concepts permeate this domain is all we can attempt here. We begin with two simple molecules and go on to touch a few areas nearer the frontiers of current research.

1 *The salt molecule* It is commonly known that salt in solution does not dissociate into two neutral atoms but into two ions, one a sodium atom with one electron missing, Na^+, the other a chlorine atom with the missing sodium electron added, Cl^-. Evidently the electron prefers being with the chlorine atom over being with the sodium atom. This immediately suggests that in an NaCl molecule the electrostatic force between the two ions may account for the bonding, which is then called *ionic*.

This hypothetical explanation of bonding in the salt molecule leaves us with some fundamental questions: Why does the electron prefer being with the chlorine atom rather than with the sodium atom, and how does this happen for sodium and chlorine and not other arbitrarily chosen pairs of atoms? By energy conservation, this ionic bond is possible if the energy *required* to strip the most loosely bound outer electron away from the neutral sodium atom (ionization energy) is less than the sum of the energy *released* when an electron joins with a neutral chlorine atom (electron affinity) and the energy *released* when the resulting positive and negative ions join together (electrostatic potential energy). Therefore, to understand the salt molecule we must look at the wave-mechanical origin of ionization energy and electron affinity and see how these properties vary through the periodic table.

For atoms beyond hydrogen, which involve three or more particles, it is not yet possible to obtain the exact form of the overall

[1] Even the classical Newtonian three-body problem of finding the motion of three bodies under their mutual gravitation has not been solved exactly in the general case. The same kind of mathematical difficulties stand in the way.

Sodium atom: one electron outside a closed shell

wave function of the atom.[1] However, approximate calculations can be made for such systems, and the resulting electron densities are not unlike, at least qualitatively, the densities one would obtain by building the atom out of the electron charge distributions calculated for the hydrogen atom (see, for example, Figs. 26-16 and 26-19). One starts with the lowest energy state and adds one charge distribution for each electron of the atom, taking proper account of the Pauli exclusion principle. Let us see how this approximate treatment leads to an understanding of the observed ionization curve shown in Fig. 26-25.

First recall that in discussing the periodic table we noted that the total charge distributions of the *rare-gas* atoms, for example, $_{10}$Ne, are spherically symmetric. The charge clouds of the individual electrons combine without cavities around the nucleus, making the net charge density in any one direction the same as in any other. This implies something very significant about any neutral atom, for example, $_{11}$Na, which has just one more electron (chemical valence $+1$) than a rare-gas atom (in the case of sodium, $_{10}$Ne). We think of the outer eleventh electron of $_{11}$Na as experiencing the electric field from the positively charged nucleus and from the 10 inner electrons, but the charge distribution of the 10 inner electrons is spherically symmetric, just like the electronic charge distribution of $_{10}$Ne. Thus the electric field of the 10 inner sodium electrons is the same as though they were all concentrated at the nucleus, there canceling out all the positive nuclear charge except an amount equal to one unit, just the charge of the hydrogen nucleus. The spherically symmetric charge distribution of the inner sodium electrons has, so to speak, *shielded* the outermost electron from $(Z-1)/Z = \frac{10}{11}$ of the nuclear charge. Consequently, the ionization energy of the sodium atom ought to be nearly equal to that of the hydrogen atom in its ground state. Actually, it is just a little smaller than that (see Fig. 26-25), because the outermost sodium electron is farther from its nucleus than the electron in hydrogen is from its nucleus.

A drastically different situation exists when one wants to ionize a rare-gas atom, for example, $_{10}$Ne. Instead of experiencing the electric field of a spherically symmetric charge distribution, the outermost electron of neon is acted on by the electric field from the neon nucleus plus an electronic charge distribution with big cavities in it. In the sodium atom, these cavities were plugged by the tenth inner electron, but now it is the tenth electron itself that is being pulled away. Thus, the inner nine electrons of the neon atom do not shield the tenth electron well. The nucleus and its nine inner electrons exert at each distance a stronger force on the outer electron than in the case of sodium. Moreover, with the cavities in the neon inner core, its outer electron can move in closer to the nucleus, at least in certain angular regions. The overall result is that the outer electron is much more strongly held in neon than in sodium. Here we have a qualitative explanation of the sharp discontinuity in the ionization curve between each rare-gas atom (helium, neon, argon, etc.) and the next atom in the periodic table (the *alkali metals,* lithium, sodium, potassium, etc.).

This same general kind of force which holds the outer electron of the rare-gas atoms so tightly also accounts for the strong electron affinity in the atoms which immediately precede the rare-gas atoms (chemical valence -1) in the periodic table (the *halogens,* fluorine, chlorine, etc.). An electron approaching a neutral halogen atom, e.g., chlorine, finds the charge distribution very similar to that of the nucleus and inner electron core of the succeeding rare-gas atom, neon. Of course, now the electric field will not be quite so strong because the halogen nucleus has one unit less of positive charge than the rare-gas atom. But still the approaching electron is not well shielded from the nucleus; furthermore, it can penetrate the cavities just like the outer electron of the rare-gas atom. In sum, the additional electron is strongly attracted to the halogen atom. As for a rare-gas atom, its electron affinity is of course very small: the nucleus is well shielded, and its charge cloud offers no cavities which the additional electron might penetrate. The net result is that the observed electron-affinity curve, Fig. 26-26, has the same sawtooth shape as the ionization-energy curve, Fig. 26-25. There are, however, two big differences: (1) the actual electron-affinity values (energies) are smaller than the ionization-energy values because the electric force is smaller, as described above; and (2) the downward jumps in the electron affinity come, as explained above, one atomic number earlier (between halogens and rare gases) than the jumps in the ionization energy (between rare gases and alkali metals).

Looking at the two curves in the region between neon and argon, along with a calculated electrostatic energy for NaCl (5.4 eV), we find the ionization energy of sodium is actually smaller than the sum of the electron affinity of chlorine and the electrostatic energy. The conditions are then met for the formation of the Na$^+$ and Cl$^-$ ions and their ionic bonding into NaCl.[1]

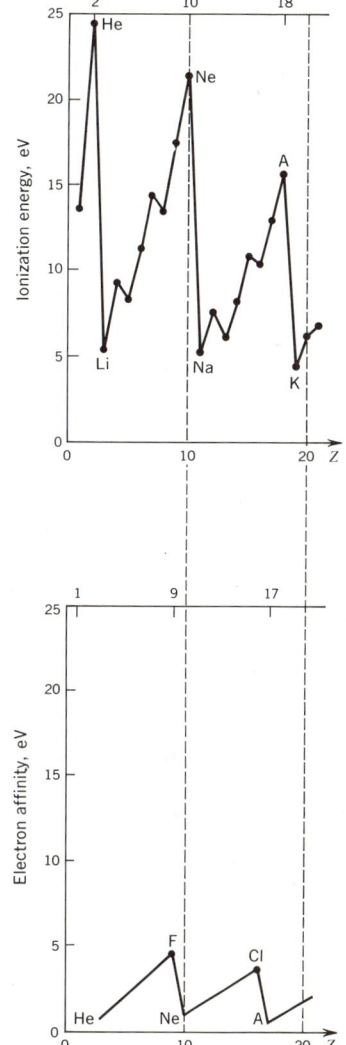

FIGURE 26-25 Ionization energies for elements as a function of their atomic number.

FIGURE 26-26 Electron-affinity energies for elements as a function of their atomic number.

[1] However, for *large* separations (electrostatic energy zero), Na + Cl is more stable than Na$^+$ + Cl$^-$; thus, the NaCl molecule in the gaseous state dissociates into atoms, not ions.

FIGURE 26-27 (*a*) A bare proton relatively far from a hydrogen atom; (*b*) at closer separation distances the electron is shared equally by the two protons, and the H_2^+ molecule is formed.

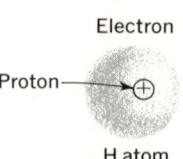

Covalent bonding illustrated by the hydrogen molecule

2 *The hydrogen molecule* If wave properties are important for an electrostatic bond, they are even more so for the bonding of a molecule like hydrogen, which splits into two *neutral atoms* rather than two ions. In ionic bonding each electron is rather closely associated with one atom or the other, but in the type of bonding found in the hydrogen molecule (called *covalent bonding*) the electrons are shared by the two atoms. To picture how this might come about, consider what happens classically when a bare proton approaches a neutral hydrogen atom. The electron and proton of the neutral atom, being very close to each other, initially experience very nearly the same force from the approaching proton. So the electron, being much less massive than the proton nucleus, moves somewhat nearer to the bare proton than the nucleus does (Fig. 26-27*a*). We say the neutral atom has become *polarized,* and because of this separation of charge in the neutral atom, the approaching proton is a little more strongly attracted by the electron than it is repelled by the nucleus. As the proton gets near the neutral atom, we have a three-particle system which must be analyzed taking full account of wave properties. In view of what our classical model has told us, it is not surprising that the approximate (there being three particles) wave-mechanical solution shows a ground state in which the electron charge density is greatest in the region between the two protons; one can attribute the stability of the molecular ion to the fact that the repulsive force on one proton arising from its interaction with the other proton at one special separation distance is balanced by the attractive force from the near but somewhat smeared-out electronic charge distribution between the two protons (see Fig. 26-27*b*). At larger separation distances the *net* force is attractive, and at smaller distances, it is repulsive.

What happens now when a second electron is added to this system to form the neutral hydrogen molecule? Here the (again approximate) wave-mechanical solution shows that the electron density in the lowest energy state of the four-particle system is one in which the charge distribution for *both* electrons is concentrated in the region between the protons (like Fig. 26-27b). This accords with our discussion of the molecular ion (there the single electron was between the protons) and with our experience with the Pauli exclusion principle (two electron charge clouds can have the same region of highest concentration if the electrons have spin angular momenta of opposite sense). The description of the covalent bond as one in which two electrons of opposite spin are shared by the two atoms is thereby consistent with the results from the wave-mechanical analysis.

3. *More complicated molecules* Atomic electrons with orbital quantum number $l = 0$ have, as we have seen, spherically symmetric charge distributions. We should not expect to be able to associate a definite direction with bonds formed via such electrons; hence we should not expect the resulting structures to exhibit different properties in different directions. However, bonds formed with electrons having asymmetrical charge distributions (having angular dependence) may indeed have definite orientations with respect to other bonds in the molecule. These directional characteristics permit the building up of elaborate three-dimensional networks of atoms, such as occur in polymers and proteins. Thus, the complicated molecular structures of chemistry and biology ultimately owe their existence to the wave nature of the electron.

In all but the simplest cases, it is largely an idealization to characterize a molecular bond as ionic or covalent. Wave-mechanical solutions commonly indicate that although an electron is being shared by two atoms, it may indeed spend much, even almost all, of its time nearer one atom than the other.

4. *The condensed states of matter* Although the general features of dense gases, liquids, and solids are well known, many related phenomena are still incompletely understood in terms of fundamental principles. Indeed, a theoretical understanding of the stability of bulk matter (why the floor holds us up) was attained only recently.[1]

One might well ask how it is possible to "solve" a system containing billions of billions of atoms. It certainly is not possible

---

[1] This achievement was reported in two papers by Freeman J. Dyson and A. Lenard, *Journal of Mathematical Physics*, March, 1967, and May, 1968.

to use the same approach or to obtain the same accuracy as in the solution of the hydrogen atom. Fortunately, in the realm of many-atom systems, simplifying models can be found that make it possible to concentrate on the more significant factors. When dealing with large numbers of atoms, the physicist is more like a cartoonist than a photographer: he intentionally sacrifices detail and emphasizes the important relationships. The theories of gases, liquids, and solids illustrate this approach.

Although there are excellent model-based theories for the solid and gaseous states of matter, there is as yet no comparable model or theory for the liquid state. This is not too hard to understand. In solids, we observe long-range order that is so dominant compared to the disordered thermal motion that a perfectly regular crystal structure makes a very good starting model. At the other extreme, in a dilute gas, as we know from Chapter 10, the chaotic thermal motion vastly predominates over the order-bringing intermolecular forces. Therefore, complete randomness constitutes a reasonable initial model. But in liquids and dense gases *both* the order from intermolecular forces and the disorder from thermal motion are important. It becomes exceedingly difficult to devise a properly schizophrenic model for such substances. Nevertheless, considerable research effort is being devoted to liquids, especially the condensation of gases into the liquid state and the transition of the liquid state into the solid. Also of particular interest is the critical point where the properties of gaseous vapor and liquid merge (Fig. 26-28).

Superconductivity

The study of the solid state is one of the most highly developed branches of modern physics. The wave properties of electrons have made clear the differences between electrical conductors and insulators; they have also provided, for the first time, an explanation for the way in which the resistance of a metal varies with temperature. Quantum theory and experiment have gone hand in hand in developing semiconductors and transistors, major components of modern electronics. And one of the most intriguing areas of research is the domain of temperatures near absolute zero, where *superconductors* seem to offer no resistance whatever to the flow of electric current. Currents in these materials have been observed to circulate around a ring for years without detectable diminution. A fascinating feature is that in pure samples of certain metals not subjected to strain the superconducting state

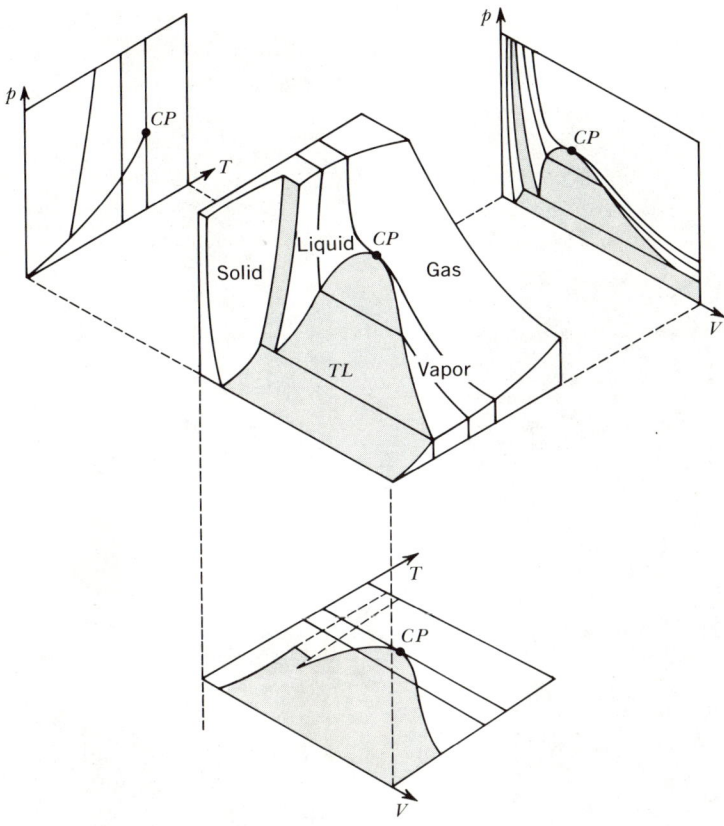

FIGURE 26-28 Phase diagram of a simple substance such as xenon showing the variation of the pressure p with changes in temperature T and volume V for a fixed number of particles. All points on the three-dimensional pVT surface, except for the shaded regions, correspond to possible equilibrium conditions for a single phase. Volumes corresponding to the shaded regions are partially occupied by one phase in coexistence with a second phase, except for points on the triple line, TL, where all three phases can coexist. At the critical point CP the liquid and vapor phases merge and become indistinguishable.

is not approached gradually but attained abruptly at a particular critical temperature.

Other materials, too, behave paradoxically at very low temperatures. Liquid helium below 2.3°K conducts heat very efficiently, much better than many metals; it also flows through very fine capillary tubes or over surfaces without needing any force to drive it. Indeed, when a cup of liquid helium is dipped from a bath, the helium runs *up* the inside walls of the cup and drips back into the bath (Fig. 26-29)!

Quantum mechanics has made substantial inroads on these *superfluids,* as well as on superconductors, but much remains for future study.

620

CHAPTER 26
QUANTUM SYSTEMS OF PARTICLES

(a)

(b)

FIGURE 26-29 The superfluidity of liquid helium: (a) liquid runs *up* and over the sides of the test tube, dripping back into the bath; (b) liquid mounts through the capillary and continues upward as a jet. (*Arthur D. Little, Inc.*)

SUMMARY The wave function ψ, which makes it possible to calculate the probability that a particle will be found in the vicinity of any given point of space, is determined from the *Schrödinger equation* much as a particle's precise, classical location is determined from basic equations, those of Newtonian mechanics.

Four quantum numbers identify the state of a hydrogen atom: the *principal quantum number n*, which specifies the atom's energy; the *orbital quantum number l*, which relates to the angular momentum of the orbiting electron; the *orbital orientation quantum number m_l*, which gives the orientation of the plane of the electron's orbital motion; and the *spin orientation quantum number m_s*, which labels the two possible orientations of the electron's spin.

The *Pauli exclusion principle* asserts that no two electrons of the same atom have the same set of four quantum numbers. The ordering of atoms according to this principle accounts for the periodic table of the elements.

PROBLEMS

26-1 Waves travel along a Slinky at a speed of 2.0 m/s. The ends of the Slinky are fixed in position, 6.0 m apart. What are the five lowest resonance frequencies of the Slinky?

26-2 Suppose that a Slinky is fixed at one end but free to move at the other, as shown in Fig. 18-15. Sketch the standing-wave patterns for the three lowest resonance frequencies.

26-3 Sketch the standing-wave pattern for waves along a circular membrane corresponding to the quantum numbers $j = 3$ and $k = 4$.

26-4 The allowed standing-wave patterns for a Slinky attached at both ends are such that the distance L between the ends of the Slinky is an integral multiple of a half-wavelength $\lambda/2$, i.e., $L = n(\lambda/2)$, where $n = 1, 2, 3, \ldots$ This relation can be derived on the following basis: if a traveling wave is not to interfere destructively with itself after being reflected once from each of the two fixed ends, the time interval Δt for a given point on the wave to travel one complete round trip must be an integral multiple of the wave's period of oscillation T, i.e., $\Delta t = nT$. Show that the first relation follows from the second one.

26-5 An atom of neon, atomic number 10, is in its lowest energy state. Give the four quantum numbers for each of the 10 electrons.

26-6 Sketch the wave function ψ as a function of the distance r from the nucleus for a hydrogen atom having zero angular momentum in the third ($n = 4$) excited state.

26-7 (a) A kettledrum struck at its center will oscillate in those standing-wave patterns for which $k = 0$ but not those modes for which $k = 1, 2, 3, \ldots$ Why? (b) Where approximately should the kettledrums be struck to excite the oscillation mode $j = 2$, $k = 1$?

26-8 Explain qualitatively, in terms of the quantum numbers and the Pauli exclusion principle, why the valence of (a) Mg is $+2$, (b) F is -1, and (c) C may be $+4$.

26-9 In Problems 18-15 and 18-16 we saw how $y = A \sin(kx - \omega t)$ represents a wave traveling to the right and $y = A \sin(kx + \omega t)$ a wave traveling to the left with $k = \omega/v$. Starting with these two mathematical representations, derive the mathematical expression for the standing waves shown in Fig. 26-1. The length of the Slinky (or string) is L, and the wave speed is v. *Hint:* For all four examples take $y = 0$ for both $x = 0$ and $x = L$. (b) What are the resonant frequencies and wavelengths?

26-10 We have mentioned the solid, liquid, and gaseous states of matter. Why do you suppose the authors of "Seven States of Matter"[1] used the number seven in the title?

[1] A Westinghouse Search Book written by scientists of the Westinghouse Research Laboratories and published by Walker and Company, New York, 1966. Chapter 1 will help you answer this question, but you may want to keep on reading!

NUCLEAR PHYSICS CHAPTER TWENTY-SEVEN

CHAPTER 27 NUCLEAR PHYSICS

In atomic physics the nucleus enters only as a massive point charge controlling the motion of electrons which surround it. So was it treated in preceding chapters. But the nucleus itself has an internal structure. It is, in fact, composed of simpler particles held together by a force enormously stronger than any encountered in the atomic domain. In this chapter we shall examine the nuclear force and the composition of nuclei, their relative stability, and changes in nuclear structure that take place spontaneously or as a result of bombardment with outside particles.

+ 27-1 THE PROTON-ELECTRON NUCLEAR MODEL

Today most of us learn early in school that the atomic nucleus consists of positively charged protons and electrically neutral neutrons, but behind that statement lie so many fundamental discoveries, both theoretical and experimental, that it is worthwhile to begin with the picture of the nucleus that prevailed till about 40 years ago.

Before 1932 it was thought that the proton, the nucleus of an ordinary hydrogen atom, was the only massive constituent common to all nuclei. It seemed as though the nucleus of each element consisted of protons, to the number required by the mass of the atom, plus however many *nuclear* electrons were needed for an overall neutral atom. For example, an atom of ordinary carbon 12 has a mass very nearly 12 times that of a proton. By definition, in fact, the mass of a neutral carbon 12 atom is taken as 12.0000 ··· in the so-called unified atomic mass scale. Since the nucleus is normally surrounded by 6 atomic electrons with total charge $-6e$, the nucleus itself must have a net charge of $+6e$. Thus, besides the 12 protons, the carbon 12 nucleus would contain $12 - 6 = 6$ electrons in order for the whole atom to be electrically neutral. The presence of these nuclear electrons would not appreciably affect the value of the atomic mass; an electron is only about 1/2,000 as massive as a proton.

Unified atomic mass scale: neutral carbon 12 atom has mass of exactly 12 u.

For any other element the argument runs the same way. Let us write X for the element's chemical symbol, Z for its atomic number, or nuclear charge, measured in units of e, and A for its atomic mass, measured in atomic mass units (u). Then, on the proton-electron model, the nucleus ^A_ZX is made up of A protons and $A - Z$ electrons. This satisfactorily explains the mass A and the net nuclear charge $Ze = A(+e) + (A - Z)(-e)$.

Nonetheless, the proton-electron model had to be discarded. Its downfall had already been presaged by the results of Rutherford's scattering experiments; in showing that Coulomb's law does not describe the interaction between an alpha particle and a nucleus when their separation is less than about 10^{-14} m, these experiments effectively determined the proton radius as $\sim 10^{-15}$ m. An electron held within 10^{-15} m by the electrostatic attraction of a neighboring proton would have an energy of the order of $k_e e^2/r \approx 1$ MeV (see [14-10]), but, according to the uncertainty principle, if an electron's extension in space is no greater than 10^{-15} m, its energy (Example 25-6) must be more than 500 MeV! This discrepancy is enough to rule out electrons as nuclear constituents.

The uncertainty principle forbids electrons in the nucleus.

If *electrons* cannot exist inside the nucleus, what is responsible for the fact that the nuclear mass number A exceeds the electric charge number Z? One might guess that the additional mass arises from the presence in the nucleus of neutral particles which are not just coupled electrons and protons. Indeed, with the discovery of the *neutron* in 1932 and with the invention of high-energy particle accelerators in the same year, a totally new understanding of nuclear structure became possible.

27-2 DISCOVERY OF THE NEUTRON

James Chadwick's discovery of the neutron is particularly valuable as an introduction to contemporary particle physics. The experimental procedures involved and the theoretical methods applied are basically the ones used today in analyzing interactions between all sorts of esoteric subnuclear particles.

Chadwick built upon the curious results of an earlier investigation by Bothe and Becker, whose setup is diagramed in Fig. 27-1. When alpha particles of about 5 MeV were emitted from a radioactive source (polonium) housed in a vacuum chamber, they struck a slab of beryllium, from which highly penetrating radiation was found to emerge. It passed without appreciable diminution through thick layers of lead and was capable of darkening photographic plates, just like visible light or x-rays. When passed through a gas, it produced pairs of positively and negatively charged ions, much as protons or alpha particles do. This penetrating radiation, however, was undeflected by the strongest magnetic fields, which implied that whatever particles composed it were electrically neutral. The problem that remained was to establish these particles as photons of zero rest mass or alternatively discover their rest mass.

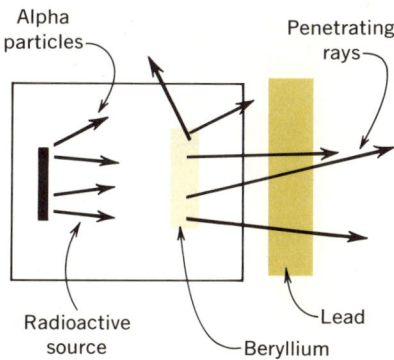

FIGURE 27-1 Arrangement for producing the penetrating rays used in the Chadwick experiment: alpha particles from a radioactive source strike beryllium, and rays capable of passing through a slab of lead emerge from it.

Dealing with electrically neutral particles is always an indirect and subtle business, since all devices for registering the presence of particles ultimately depend upon an unbalance in electric charge. For example, a photon can be detected and register a "count" in an instrument if it collides with an atom and dislodges an electron through the photoelectric or Compton effect; the electric charge of the electron set in motion (and that of the remaining ion) are responsible for the electrical unbalance. Even when a visible-light photon is detected directly by the eye, it is the electric unbalance produced at the retina which is transmitted through electrical conductors (nerves) to the brain. Similarly, to detect and study the properties of a neutral material particle one must find out how such a particle influences charged particles.

In another set of experiments, performed by Frédéric and Irène Joliot-Curie, the mysterious penetrating radiation was scattered from two separate targets, whose constituent particles differed in mass by a factor of 14 (Fig. 27-2). The target particles which are struck head on receive maximum kinetic energy and recoil in the forward direction. This fact was seized on by Chadwick, who saw that it permitted application of the conservation laws in their simple one-dimensional forms to determine the mass of the unknown neutral particles.

Chadwick counted the total number of ion pairs produced by a recoiling particle in coming to rest. Since the energy required to produce a single ion pair is known, this measurement yields the initial kinetic energy of the recoiling particle; and since the particle's mass is also known, one can compute its initial speed. The results of Chadwick's measurements are summarized in Table 27-1. The kinetic energies and speeds are uncertain by about 10 percent because of rather primitive instrumentation.

Chadwick's analysis proceeded as follows. First of all, one may assume that the incoming neutral particles have a *single* kinetic energy because the alpha particles responsible for producing them in the beryllium slab have a single kinetic energy of around 5 MeV. Next, the possibility of the neutral particles being photons can be ruled out. To do this, suppose, on the contrary, that the neutral particles are photons. We know that when a photon hits a free particle initially

What are the incident particles, given the masses and kinetic energies of the recoiling target particles?

TABLE 27-1

Recoiling Target Particle	Mass, u	Maximum Recoil Kinetic Energy, MeV	Maximum Recoil Speed, $\times 10^7$ m/s
Proton	1	5.7	3.3
Nitrogen nucleus	14	1.4	0.47

at rest in a Compton collision, the struck particle recoils and a second photon is created and leaves the collision site with less energy than the incident photon. The incident photon transfers maximum energy to the struck particle if the Compton collision is head on; then the particle recoils in the forward direction, and the scattered photon travels backward. (Even though the struck protons and nitrogen 14 nuclei are not perfectly free but are bound to neighboring charged particles, this energy of binding, at most a few electron volts, is so small compared to their recoil energies that we may treat the target particles as free and initially at rest.)

Table 27-1 shows that the speed of the recoiling particles is appreciably less than that of light, so that we can properly take the recoiling particle's momentum and energy to be given by the classical expressions mv and $\frac{1}{2}mv^2$, respectively. Denoting the incident photon by the subscript i and the back-scattered one by s, we can write

Energy conservation: $\quad h\nu_i = \frac{1}{2}mv^2 + h\nu_s$

Momentum conservation: $\dfrac{h\nu_i}{c} = mv - \dfrac{h\nu_s}{c}$

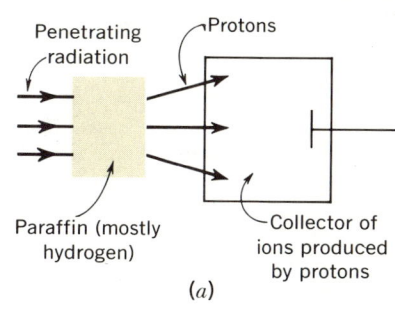

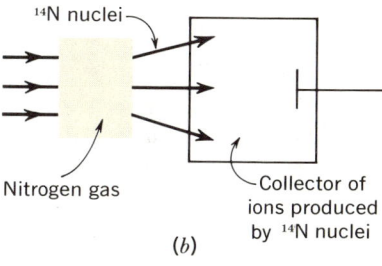

FIGURE 27-2 The penetrating radiation (*a*) incident upon protons, which are collected in a detecting device, and (*b*) incident upon nuclei of nitrogen 14.

Eliminating $h\nu_s$ from these two equations yields

$h\nu_i = \frac{1}{2}(\frac{1}{2}mv^2 + mvc) = \frac{1}{2}[\frac{1}{2}mv^2 + \sqrt{2(\frac{1}{2}mv^2)(mc^2)}]$

or, with $E_k = \frac{1}{2}mv^2$ and $E_0 = mc^2$,

$h\nu_i = \frac{1}{2}(E_k + \sqrt{2E_k E_0})$ [27-1]

It is easy to show that [27-1], which applies if the incident particles really are photons, is incompatible with the data in Table 27-1. Substituting into [27-1] the values of E_k and E_0 for a recoiling proton, we find $h\nu_i = 55$ MeV, whereas for a nitrogen nucleus we find $h\nu_i = 100$ Mev! Two different values for what should be the same number and both too large.

Energy and momentum conservation show that the incident particles are not photons.

Therefore, knowing that the unknown neutral particles must have a positive rest mass, we can rewrite the conservation equations and use them in combination with Table 27-1 to obtain that rest mass.

Let us anticipate the naming of the particles as neutrons by using the subscript n. If such particles with initial speed v_n and a final speed (after collision) of v'_n strike target particles of mass m initially at rest in a head-on elastic collision, we have

Energy conservation: $\quad \frac{1}{2}m_n v_n^2 = \frac{1}{2}mv^2 + \frac{1}{2}m_n v'^2_n$

Momentum conservation: $m_n v_n = mv - m_n v'_n$

where v is again the recoil speed of the target particle. The masses and kinetic energies, or speeds, of the two types of target particle are known by experiment; the masses and the speeds of the incident particles before and after collision are not. We wish to compute the penetrating particle's mass m_n in terms of m and v.

We begin by eliminating v' from the two equations above. After some algebra we find that

$$[27\text{-}2] \quad v_n = \frac{m + m_n}{2m_n} v$$

For protons as target particles, $m = 1$; their speed of recoil is designated v_1. Equation [27-2] then becomes

$$v_n = \frac{1 + m_n}{2m_n} v_1$$

With nitrogen nuclei as target particles the mass is $m = 14$, and the final speed is $v = v_{14}$. We then have the second equation

$$v_n = \frac{14 + m_n}{2m_n} v_{14}$$

Dividing this equation by the one above it and using the speeds listed in Table 27-1 yields

$$1 = \frac{14 + m_n}{1 + m_n} \frac{v_{14}}{v_1}$$

or

$$\frac{14 + m_n}{1 + m_n} = \frac{v_1}{v_{14}} = \frac{3.3 \times 10^7 \text{ m/s}}{0.47 \times 10^7 \text{ m/s}} = 7.0$$

Neutron mass approximately 1 u

Solving this last equation for m_n gives $m_n \approx 1$ u. We recall that the measured quantities in Chadwick's experiment were uncertain by at least 10 percent. Therefore, at least to the precision which this experiment allows, we can write $m_n = 1$ u.

In this manner, from the behavior of known, *charged* particles, the mass of new uncharged particle was found. Later and more precise measurements show that the neutron and proton masses are the same to within 1 part in 10^3. Thus, when a neutron strikes a stationary proton in a perfectly elastic collision and the proton thereby is given a kinetic energy of 5.7 MeV (Table 27-1), the neutron is brought to rest and all its initial kinetic energy is transferred to the proton. Therefore, the neutrons produced in the Chadwick experiment had a kinetic energy initially of 5.7 MeV, which is of the same order

TABLE 27-2

	Symbol	Mass, u	Rest Energy, MeV	Electric Charge
Proton	$_1^1p$	1.0072766	938.256	$+e$
Neutron	$_0^1n$	1.0086654	939.550	0

of magnitude as the kinetic energy of the alpha particles striking the beryllium target (Fig. 27-1). The properties of the neutron are compared to those of the proton in Table 27-2.

By itself the Chadwick experiment does not establish that the neutrons came from *nuclei,* but a variety of additional experiments point to that conclusion. We now accept neutrons as basic constituents, along with protons, of all nuclei heavier than hydrogen. The "missing" $A - Z$ mass units in the nucleus $_Z^A X$ are now accounted for: the nucleus is made up of Z protons and $A - Z$ neutrons (and not of A protons and $A - Z$ electrons, as the proton-electron model had it).

27-3 THE NUCLEAR FORCE

The nuclear force acts between nucleons within about 10^{-15} m.

We have already estimated the size of the nucleus from the departure of the force from Coulomb's law at separation distances of less than about 10^{-14} m (which is about 1/10,000 the size of a typical atom). In fact, at distances under 10^{-15} m two protons strongly *attract* each other. This net attractive force, which is also quite distinct from the relatively feeble force of gravity, is known simply as the *nuclear force;* we also sometimes speak of it as the *strong interaction.*

The characteristics of the nuclear force between protons can be gauged in scattering experiments of the type discussed in Section 27-1. By shooting energetic protons at protons at rest in a target and analyzing the number emerging in various directions, one can deduce the force acting between pairs of particles. In similar fashion, by directing a beam of neutrons at a target consisting mainly of hydrogen atoms, hence of protons, one can establish that a neutron strongly *attracts* a proton, but only for separation distances less than about 10^{-15} m. At larger distances the neutron-proton force is zero. Figure 27-3 shows the potential energy (and therefore, through its slope, the force) between a pair of protons and between a neutron and a proton.

It is also known, although less directly, that a neutron attracts a neutron through a strong force, which again operates only within a range of approximately 10^{-15} m. In short, the nuclear force acts

FIGURE 27-3 (a) Potential energy between a pair of protons as a function of their separation distance r. The interaction is that of the Coulomb force for $r \gtrsim 10^{-15}$ m and that of the nuclear force for lesser distances. (b) Potential energy between a proton and neutron as a function of their separation distance r. The force between the particles is zero for $r \gtrsim 10^{-15}$ m and that of the attractive nuclear force for lesser distances.

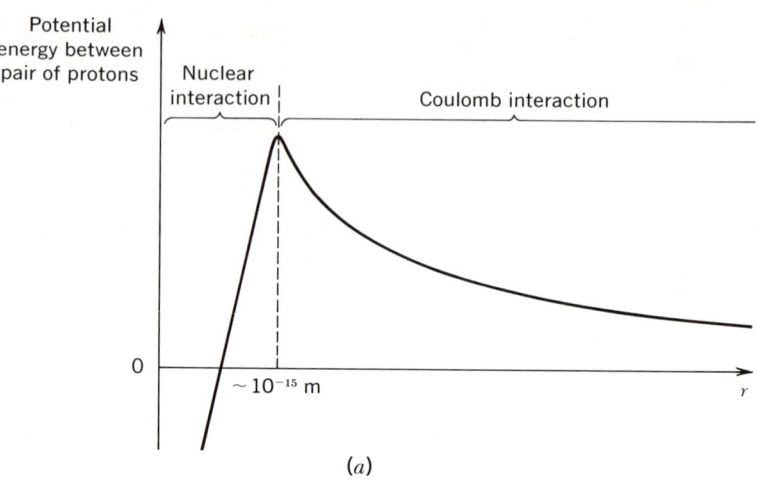

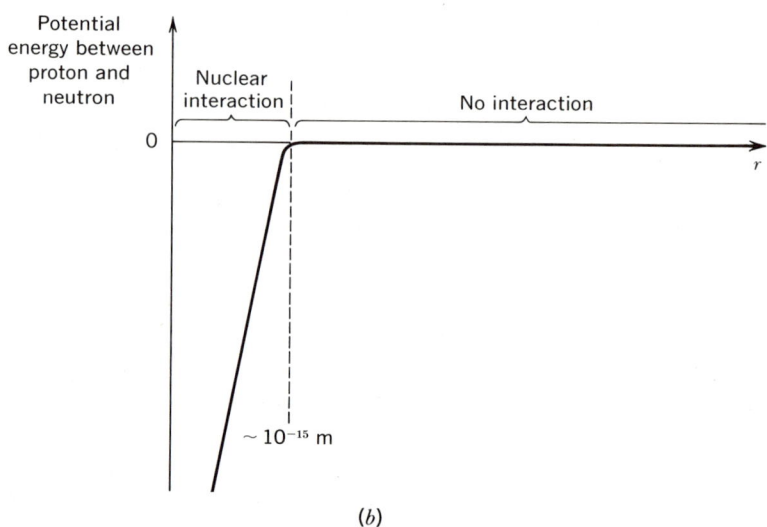

between any pair of *nucleons*, where by a nucleon is meant either a proton or neutron. Not only does the nuclear force, or strong interaction, hold the nucleus together, but it is the principal interaction between many other elementary particles, as we shall see in Chapter 28.

27-4 NUCLEAR BINDING ENERGY

We wish to consider the binding of nucleons within nuclei. Let us first see how particles are bound in a familiar system, the hydrogen atom. Suppose that a proton and electron are initially far separated and at rest. They attract each other through the Coulomb force, come together, and form a hydrogen atom. But the hydrogen atom is in its lowest, or ground, state only after a photon of 13.6 eV has been emitted (see Section 24-3). The process can be written

$$p + e \rightarrow {}_1^1\text{H} + h\nu \text{ (13.6 eV)} \quad [27\text{-}3]$$

We conclude that the hydrogen *atom* in its most stable configuration has a *binding energy* of 13.6 eV, since that amount of energy is given off in the process of assembly. Because the process begins and ends with particles at rest, the 13.6-eV decrease in the system's energy must show up as a decrease in rest mass: the rest masses of the separated proton and electron must exceed the rest mass of the hydrogen atom in its ground state by $\Delta m = \Delta E/c^2 = 13.6 \text{ eV}/c^2$. In principle, then, one could measure the energy of a hydrogen atom in its ground state, or the energy of the photon emitted, simply by comparing the sum of the rest masses of a proton and an electron with the rest mass of a hydrogen atom. In practice, this is not possible since the fractional difference in mass $\Delta m/m$, where m is the hydrogen atom mass, is

Binding energies and mass differences for atomic and nuclear systems compared

$$\frac{\Delta m}{m} = \frac{13.6 \text{ eV}/c^2}{(1.008 \text{ u})[931.5 \text{ MeV}/(\text{u})(c^2)]} \approx 2 \times 10^{-8}$$

a mere 2 parts in 10^8.

Now consider the analogous bound *nuclear* system. A proton and neutron attract each other by the strong nuclear force. Bound together, they form a *deuteron*, ${}_1^2d$, the nucleus of heavy hydrogen. Again, the deuteron will be in its ground state only after the amalgamated proton and neutron emit a photon, here of 2.2 MeV. Symbolically,

$${}_0^1 n + {}_1^1 p \rightarrow {}_1^2 d + h\nu \text{ (2.2 MeV)} \quad [27\text{-}4]$$

Notice that the binding energy, 2.2 MeV, and the associated loss in rest mass, $\Delta m = 2.2 \text{ MeV}/c^2$, are of the order of a million times greater than they are for the hydrogen atom. This large factor reflects the fact that the nuclear force is strong compared to the Coulomb force.

Because tables giving the (rest) masses of electrically neutral *atoms* are readily available, it is common practice to replace the bare nuclei in reactions such as [27-4] by the corresponding neutral atoms. Since any such replacement involves "adding" the same number of electrons to both sides of the reaction equation (charge is conserved!), the original mass balance is not disturbed.[1] It follows that in [27-4] we can replace the proton ${}_1^1 p$ on the left by an ordinary hydrogen atom ${}_1^1\text{H}$ and the deuteron ${}_1^2 d$ on the right by an atom of heavy hydrogen ${}_1^2\text{H}$ (*deuterium*), giving

[1] Except, of course, for the minute mass difference corresponding to atomic binding.

[27-5] $^1_0n + {}^1_1H \rightarrow {}^2_1H + h\nu$ (2.2 MeV)

We can now use [27-5] to make a precise determination of the binding energy of the deuteron in its ground state directly from the atomic rest masses.

Separate neutron: ^1_0n = 1.0086654 u
Separate hydrogen atom: ^1_1H = 1.0078252 u
$^1_0n + {}^1_1H$ = 2.0164906 u
Deuterium atom in ground state: ^2_1H = 2.0141022 u
Δm = 0.0023884 u

$\Delta E = \Delta mc^2 = (0.0023884 \text{ u})c^2[931.5 \text{ MeV}/(u)(c^2)] = 2.225$ MeV

We know that when a 13.6-eV photon is absorbed by a hydrogen atom in its ground state, the atom is ionized into a separated proton and electron. Symbolically,

$h\nu$ (13.6 eV) + $^1_1H \rightarrow p + e$

which is [27-3] with the arrow reversed. We must add 13.6 eV to the bound system to separate, or unbind, it into its component parts. This is the basis of an equivalent definition of *binding energy*.

By the same token, when a photon of at least 2.225 MeV is absorbed by a deuteron, the deuteron can be separated into an unbound proton and neutron. The relation is written

$h\nu$ (2.225 MeV) + $^2_1d \rightarrow {}^1_0n + {}^1_1p$

again the reverse of the amalgamation process, [27-4].

As the simple example of the deuteron shows, it is a straightforward matter to find the energy that must be added to a system of particles initially bound together in order to separate the system into its component parts. One simply compares the mass of the aggregate with the sum of the masses of the separate constituents. Unless the amalgamated particles together have less mass than the same particles separated, the particles are not bound. Consider the nucleus $^{12}_6C$. How much energy must be added to this nucleus if it is to be decomposed into 12 separate nucleons, 6 neutrons and 6 protons?

Binding energy of nucleus computed from masses of constituent particles

Six ^1_1H atoms: 6(1.007825 u)
Six ^1_0n: 6(1.008665 u)
Total *separate* masses: 12.098940 u
Carbon 12 atom: 12.000000 u
Δm = 0.098940 u

$\Delta E = \Delta mc^2 = (0.0989 \text{ u})c^2[931.5 \text{ MeV}/(u)(c^2)] = 92.16$ MeV

It takes 92.16 MeV to rip apart a $^{12}_{6}$C nucleus, or an average energy for each of the 12 particles of $(92.16 \text{ MeV})/12 = 7.68$ MeV. Comparing this figure with the binding energy per nucleon of the deuteron, $(2.22 \text{ MeV})/2 = 1.11$ MeV, we see that although both nuclei are stable, carbon is the stabler.

Carrying out the same sort of computation for all the stable nuclei gives the curve in Fig. 27-4, where the binding energy per nucleon is plotted against the total number of nucleons in the nucleus (the mass number A). The curve rises sharply with increasing values of A, reaches a broad maximum, and then falls off slightly. This implies that in the lightest nuclei the nucleons are relatively loosely bound, that they are most strongly bound together for nuclei of moderate size, and that the binding between nucleons then decreases in going to the very heaviest nuclei.

A simple interpretation of this behavior comes from picturing the nucleons as marbles packed together in a spherical array. The larger the sphere, the smaller the proportion of marbles which are "on the surface" and hence incompletely surrounded and attracted by other marbles. Therefore, the average binding energy should increase with the number of marbles, i.e., with A. There is, however, a counter-effect, due to the Coulomb force between protons: the more protons we crowd together, the more they repel each other and decrease the

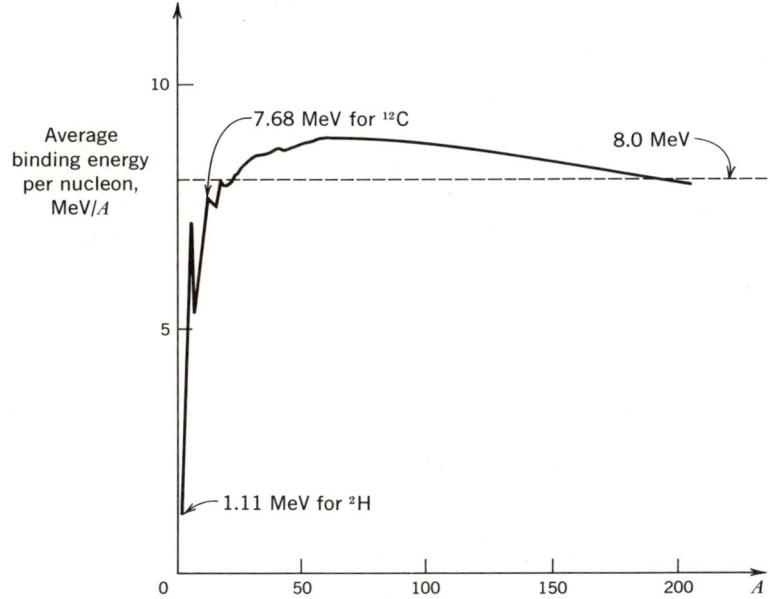

FIGURE 27-4 Binding energy per nucleon plotted against the total number A of nucleons in a stable nucleus.

average binding energy. Thus, we explain the shape of Fig. 27-4 by saying that surface effects predominate at the low-A end and that the repulsion predominates for high A. Indeed, it is this electrostatic repulsion that sets a limit to the size of nuclei that can exist as stable systems.

27-5 THE CONSERVATION LAWS IN RADIOACTIVE DECAY

In atomic physics, the emission of light by an excited atom may properly be described as a decay process. In this instance, one "particle," the atom in the excited state, decays into two particles, the atom in the lower-energy state and the emitted photon. Certain nuclei, too, can undergo *radioactive decay;* the emitted particles may be helium nuclei (*alpha decay*), electrons (*beta decay*), or photons (*gamma decay*). Whatever the decay process, it must obey the conservation laws. At the very least, the total rest mass of the decay products must be *smaller than* the rest mass of the parent nucleus; otherwise we would be dealing with a stable nucleus that could not decay in the first place. In this section we shall examine the modes of radioactive decay permitted under the conservation laws.

Gamma decay is analogous to atomic radiation. Like an atom, a nucleus has a number of quantized energy states, typically with energy differences measured in keV or MeV, as shown in Fig. 27-5. In gamma decay, the nucleus jumps from one energy level, E_2, to a lower level, E_1, with the emission of a photon (called a *gamma ray*) of energy $h\nu$. Thus, energy conservation requires that

$$E_2 - E_1 = h\nu$$

in which we have ignored, because by comparison it is very small, the kinetic energy of the recoiling nucleus after decay.

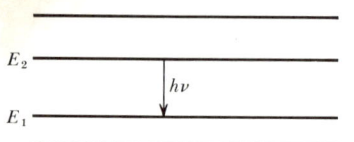

Gamma decay: photon emitted from nucleus

FIGURE 27-5 In gamma decay a nucleus spontaneously makes a transition from energy level E_2 to E_1, emitting a photon of energy $h\nu$.

The gamma emission process is shown schematically in Fig. 27-6. A free nucleus initially at rest spontaneously explodes into a photon traveling in one direction and the nucleus in the lower-energy state (the *daughter nucleus*) traveling in the opposite direction. The system's total momentum is zero before decay and, by the law of momentum conservation, must remain zero after decay. Therefore, the two particles move in opposite directions with momenta of the same magnitude. Although the photon and recoiling daughter nucleus have precisely the same momentum magnitude, the kinetic energy of the recoiling nucleus is trivially small because its mass is very large compared to the mass $E/c^2 = h\nu/c^2$ of the photon, just as the kinetic energy of the recoiling massive earth is minute compared to the kinetic energy of a ball thrown from its surface.

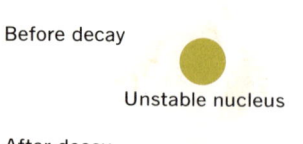

FIGURE 27-6 Gamma decay: an unstable nucleus explodes into a daughter nucleus and photon traveling in opposite directions.

An example of alpha decay is

$$^{224}_{88}\text{Ra} \rightarrow {}^{220}_{86}\text{Rn} + {}^{4}_{2}\text{He}$$

where a nucleus of radium 224 has emitted a helium 4 nucleus, or *alpha particle,* leaving a nucleus of radon 220, a different chemical element. Since the alpha particle consists of 2 protons and 2 neutrons, the atomic number of the parent nucleus is reduced by 2 (from 88 to 86) and the number of its nucleons is reduced by 4 (from 224 to 220). Thus, the conservation of electric charge is reflected in the fact that the subscripts to the chemical symbols balance on the two sides of the decay relation above. The superscripts also balance, indicating that the total number of protons and neutrons is unaltered in the decay process. This *conservation of nucleons* is in fact a general law, obeyed in all nuclear reactions. We will see in Section 28-4 that the law does *not* require that protons and neutrons be separately conserved (although that has been the case in the reactions considered so far).

In addition, the conservation laws of energy and of momentum must be obeyed in alpha decay. Energy conservation requires that the total energy of the parent nucleus equal the total energy (rest energy and kinetic energy) of the daughter nucleus and alpha particle. Thus, the decay is energetically allowed only if the total rest mass of the parent nucleus exceeds the sum of the rest masses of the daughter nucleus and alpha particle, the excess mass corresponding to the kinetic energy of the two outgoing particles.

The law of momentum conservation implies that if the parent nucleus is initially at rest and has zero momentum, the two particles must leave the decay site in opposite directions with the same magnitude of momentum (Fig. 27-7). The mass of the daughter nucleus is far greater than that of the alpha particle (220 compared to 4); consequently, the momenta being equal, a very large fraction of the total kinetic energy is carried by the light alpha particle. To summarize, the conservation of both energy and of momentum in the decay of one unstable particle into two other particles implies that the energy released as kinetic energy is shared between the two outgoing particles in such a way that the total vector momentum of the particles is zero. For a given amount of released energy there is, then, a *single* kinetic energy for the alpha particles and also a single, but much smaller, kinetic energy for the daughter nuclei. A plot of the number of alpha particles against the alpha-particle kinetic energy is as shown in Fig. 27-8. It consists of a single sharp peak, corresponding to the unique kinetic energy released in the decay process. The

Alpha decay: helium nucleus emitted from nucleus

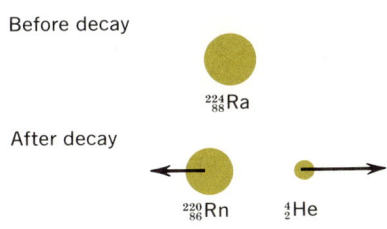

FIGURE 27-7 Alpha decay: an unstable nucleus $^{224}_{88}$Ra explodes into an alpha particle ($^{4}_{2}$He) and the daughter nucleus ($^{220}_{86}$Rn) traveling in opposite directions.

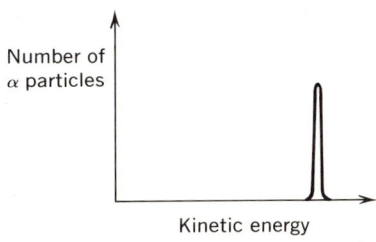

FIGURE 27-8 Plot of the number of alpha particles emitted against the energy of each, a single sharp peak.

Beta decay: electron or positron created and emitted from nucleus

situation here is like that in gamma decay, which also involves *two* final particles; there photons emerge with a single energy.

Beta radioactive decay was of momentous importance for the theory of the nucleus. One example is the spontaneous decomposition of an unstable boron 12 nucleus. The directly observed decay products are a nucleus of stable carbon 12 and an electron, or *beta particle*. We cautiously write

$$^{12}_{5}\text{B} \xrightarrow{?} {}^{12}_{6}\text{C} + {}^{0}_{-1}e$$

If it turns out that the proposed reaction satisfies *all* the conservation laws, we can confidently remove the question mark.

The electron carries a subscript -1, corresponding to its unit negative charge; the superscript 0 indicates that its rest mass is negligible in comparison to that of a nucleon. The form of the decay relation demonstrates that electric charge and nucleon number are conserved. It also shows that one of the seven neutrons in the $^{12}_{5}\text{B}$ nucleus has been changed into a proton to form the stable nucleus $^{12}_{6}\text{C}$. At the same time an electron is *created* and emitted. (As we saw at the beginning of this chapter, the uncertainty principle precludes the existence of an electron as a nuclear constituent. When an electron is emitted in beta decay, the electron is created the instant before it is emitted and simultaneously a neutron is transformed into a proton.)

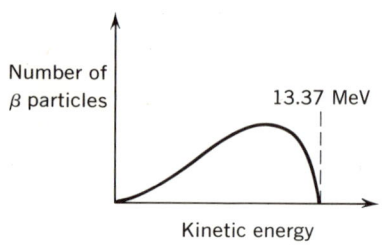

FIGURE 27-9 Plot of the number of beta particles emitted against the kinetic energy of each, a continuous distribution with an upper limit.

Next, energy and momentum must be conserved. From a table of atomic masses we find that the rest mass of the parent boron nucleus exceeds the net rest mass of the decay products by 0.01435 u. This means that (0.01435 u)(931.5 MeV/u) = 13.37 MeV of kinetic energy must be shared between the outgoing particles. Because the electron is so much lighter than the carbon nucleus, the latter should acquire essentially no kinetic energy (the argument here is the same as that used in the discussion of alpha decay). That is, beta decay should be characterized by a single kinetic energy, 13.37 MeV, for the emergent electrons.

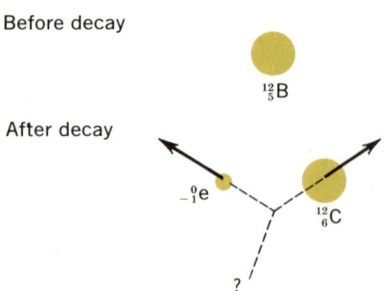

FIGURE 27-10 Beta decay: an unstable nucleus ($^{12}_{5}$B) explodes into an electron ($^{0}_{-1}e$) and the daughter nucleus ($^{12}_{6}$C), which do not travel in opposite directions; a third particle(?) is required to conserve momentum.

As actually observed the electrons do *not* emerge with a single kinetic energy; instead, the energies are distributed continuously over a wide range whose *upper limit* is 13.37 MeV, as shown in Fig. 27-9. It would appear that energy is *not* conserved! Also, experiments show that the emitted electron and the recoiling daughter nucleus do *not* move in opposite directions but as shown in Fig. 27-10. Apparently, momentum is not conserved either! Furthermore, by comparing the total angular momentum of the system before and after the decay, it can be established that angular momentum is not conserved.

One way out of this dilemma is to say that the conservation of energy, of momentum, and of angular momentum simply do not hold for beta decay. Another way out, suggested by Pauli in 1930, is to assume that a *third* particle, called the *neutrino*, is actually created and emitted in the beta-decay process.

A neutrino also created and emitted in beta decay

At first glance it is hard to decide which solution is the more desperate: abandoning the conservation laws, or inventing an exotic new particle to preserve them. As someone once put it, the neutrino is as close to nothing as something can be. It has zero electric charge (in the reaction we have been considering the beta particle and the carbon nucleus account for all the charge of the boron nucleus), and it has zero rest mass and zero rest energy (since the fastest beta particles do, in fact, have just the amount of energy required by energy conservation). Having no rest mass, the neutrino must travel at the unique speed c; yet it is distinct from a photon, which would surely have been observed if produced in beta decay. See Table 27-3.

Rest mass	$m_0 = 0$
Rest energy	$E_0 = m_0 c^2 = 0$
Electric charge	$q = 0$
Speed	$v = c$

TABLE 27-3 Neutrino Properties

Yet neutrinos really exist (they were finally detected experimentally in 1956), and beta decay can therefore be made consistent with all the conservation laws. It is the elusive neutrino that carries off the "missing" energy; and, with *three* particles involved, there are a variety of ways in which the released kinetic energy can be shared so as to conserve total momentum. Furthermore, neutrinos possess spin angular momentum (see Section 7-2), whereby angular momentum is also conserved in beta decay. As a matter of fact, two varieties of the new particle exist; one, called simply the neutrino, is denoted by the Greek letter ν, and the other, called the *antineutrino*, is denoted by $\bar{\nu}$. It turns out that it is the antineutrino that figures in the decay of $^{12}_{5}\text{B}$. Thus we finally have the true reaction equation

$$^{12}_{5}\text{B} \rightarrow {}^{12}_{6}\text{C} + {}^{0}_{-1}e + {}^{0}_{0}\bar{\nu}$$

Two other types of beta decay occur. Called *positron decay* and *electron capture*, they involve the conversion of a nuclear proton into a nuclear neutron. An example of beta decay with the emission of a positron is

$$^{12}_{7}\text{N} \rightarrow {}^{12}_{6}\text{C} + {}^{0}_{+1}e + {}^{0}_{0}\nu$$

The positron, symbolized $_{+1}^{0}e$, is identical with an electron except that its electric charge is positive. It is said to be the *antiparticle* of an electron; its properties, as well as those of other types of antiparticles, are discussed in Chapter 28.

In the process of beta decay by electron capture, one of the *atomic* electrons close to the unstable nucleus combines with a nuclear proton to yield a neutron. The only particle emerging from the decay is a neutrino (together with the slightly recoiling daughter nucleus). An example is electron capture in radioactive beryllium 7:

$$_{-1}^{0}e + {}_{4}^{7}\text{Be} \rightarrow {}_{3}^{7}\text{Li} + {}_{0}^{0}\nu$$

In all types of beta decay the nucleus changes its charge without changing the total numbers of nucleons.

+ 27-6 THE RADIOACTIVE-DECAY LAW

In the last section we considered *whether* a nucleus or even a single particle is unstable and thus subject to one or another mode of decay. The *possibilities* included any decay process not ruled out by violation of a conservation law. A second question remains: *When* will a given particle decay if it is known to be unstable? Such a question can be answered only in terms of *probabilities*. We cannot hope to predict that nucleus number 2001 definitely will or definitely will not decay in the next hour, but we can estimate the *chances* that it will decay in that period.

These chances for survival of an individual nucleus can be inferred from the behavior of a macroscopic sample of a radioactive substance, which contains billions of unstable nuclei. On the one hand, the rate at which a sample turns into inert material is a reflection of the life expectancy of an individual nucleus: the shorter that expectancy, the faster the sample decays. On the other hand, because so many particles are involved, the decay of the sample should be capable of a definite, nonrandom description. We recall, in comparison, the case of an enclosed gas, which for all practical purposes exerts a fixed pressure on the container walls, even though the individual molecular impulses are random.

Suppose, then, that we measure the number of particles emitted by a radioactive substance as time goes on. We might, for example, use as a detector a fluorescent screen and count the number of flashes produced in a 1-min interval every hour. The number of particles radiated per unit time by a collection of unstable nuclei is known as its *activity*. If we plot the activity as a function of time, the resulting graph is like that of Fig. 27-11.

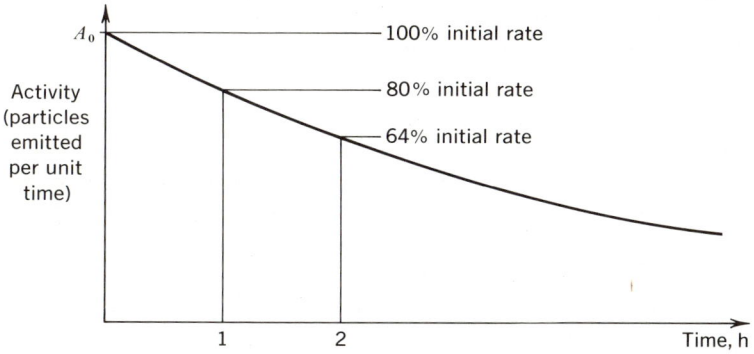

FIGURE 27-11 The activity of a radioactive material plotted as a function of time; the activity decreases by the same factor over all equal time intervals.

A characteristic feature of Fig. 27-11 and all other decay curves from unstable species is that the *activity decreases by the same factor over all equal time intervals*. Thus, if the activity drops to 80 percent of its initial value after 1 h has elapsed, the activity is found to be down by another factor of 0.80 after a second hour has elapsed, or by a net factor of $(0.80)(0.80) = 0.64$ over the 2-h period. More generally, after n 1-h intervals the activity has dropped by a factor of $(0.80)^n$ from the initial value.

Radioactive decay: decrease by same factor in activity over equal time intervals

It is easy to demonstrate that the observed decrease in activity by the same factor over all equal time intervals implies that the activity A_t at any time t is related to the initial activity A_0 at $t = 0$ through the equation

$$A_t = A_0 e^{-\lambda t} \qquad [27\text{-}6]$$

where λ is the *decay constant* of the particular species of unstable nuclei. The activity decreases exponentially with time; the larger the λ, the faster the decay. After a first time interval Δt the activity has become A_t (at $t = \Delta t$) $= A_0 e^{-\lambda \Delta t}$, where in Fig. 27-11 the quantity $e^{-\lambda \Delta t}$ is equal to 80 percent at $\Delta t = 1$ h. After a second interval, also of duration Δt, [27-6] gives the activity as A_t (at $t = 2\Delta t$) $= A_0 e^{-2\lambda \Delta t}$ $= A_0(e^{-\lambda \Delta t})(e^{-\lambda \Delta t})$. If each factor $e^{-\lambda \Delta t}$ is, say, 0.80, then $e^{-2\lambda \Delta t}$ must be 0.64.

Not only does the *activity* of decaying unstable particles decrease exponentially with time, so too does the *number* of surviving unstable particles. One simply assumes that the activity is determined by a property of the individual particles. Clearly, then, if the number of unstable particles at some one time is cut to half, the activity at that instant is also cut to half. In other words, the number of surviving

640

CHAPTER 27
NUCLEAR
PHYSICS

Law of radioactive decay: exponential decrease in activity or in number of undecayed particles with time

unstable particles N_t at time t must be proportional to the activity A_t at that time. Therefore, the relation giving the time dependence of N_t is of the form

[27-7] $$N_t = N_0 e^{-\lambda t}$$

where N_0 is the initial number (at $t = 0$) of unstable particles and the rate of decay is again controlled by the decay constant λ. As will be shown in Example 27-1, the number of undecayed particles and the activity at the same time are related by $A_t = \lambda N_t$ and $A_0 = \lambda N_0$.

Equation [27-7], giving the number of surviving unstable particles, is the *law of radioactive decay*. This single remarkably concise formula describes the decay of every species of unstable particle, each species having its own value of the *decay constant* λ. It is evident from the form of [27-7] that, just like the activity, *the number of undecayed particles decreases by the same factor over all equal time intervals.* A graph of N_t vs. t (Fig. 27-12) has exactly the same shape as the activity curve (Fig. 27-11).

We are now prepared to descend to the level of the individual radioactive particle and to infer the probability that a particular undecayed particle will survive decay over the next t s. Rewritten as

$$\frac{N_t}{N_0} = e^{-\lambda t}$$

the law of radioactive decay tells us that out of a *large number* N_0 of undecayed particles present at $t = 0$, a fraction $e^{-\lambda t}$ are still undecayed after t s has elapsed. Since all the N_0 particles are identical and indistinguishable, we can rightly conclude that $e^{-\lambda t}$ represents

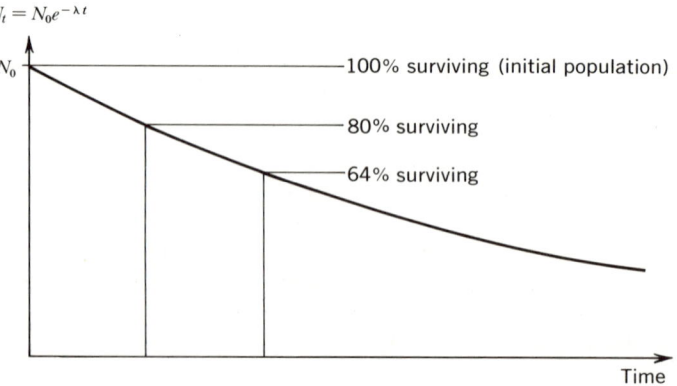

FIGURE 27-12 Number N_t of undecayed particles as a function of the time t; the number of surviving particles decreases by the same factor over all equal time intervals.

the *probability* that any single particle will survive for at least t s. (Compare "3,000 of this year's 9,000 Eureka automobiles have brake defects. The probability that my new Eureka has a brake defect is 1/3.") Thus we have for the underlying law governing the decay of individual unstable particles:

Probability an undecayed particle will survive
for at least t more seconds $= e^{-\lambda t}$ [27-8]

where λ is the decay constant of the species of particle in question.

The transition made above from a mere *fraction to a probability* is trickier than might appear; it may be helpful to view the matter the other way around. Suppose that a certain kind of unstable particle had a probability of 1/6 for surviving 1 h. Then if we observed a small collection, say $N_0 = 100$, of these particles, we should expect around 16 survivors after 1 h. Perhaps 14, perhaps 19. Other outcomes are, of course, *possible*. It is even conceivable that all 100 particles should survive, but the odds are enormously against it. With $N_0 = 1,000$, the chances grow even remoter of observing a surviving fraction significantly different from 1/6. Finally, for a typical sample of, say, $N_0 = 10^{15}$ particles, for all practical purposes we may assume that precisely 1/6 of the particles remain undecayed after 1 h.

This way of looking at the decay process, from the vantage point of the individual particles and their random lifespans, shows that the law of radioactive decay [27-7] is actually an instance of what is sometimes called *the law of averages*. Hence our insistence that N_0 be large in applications of [27-7].

The chances for survival of an individual particle and the decay rate of a macroscopic sample are both controlled by the size of the decay constant λ. Another measure of particle survival is the half-life $T_{1/2}$. By definition, the half-life of an unstable particle is that period of time over which the particle has a fifty-fifty chance of surviving. According to [27-8], $T_{1/2}$ and λ are connected by

Half-life defined

$$\tfrac{1}{2} = e^{-\lambda T_{1/2}}$$
$$-\ln 2 = -\lambda T_{1/2}$$
$$T_{1/2} = \frac{\ln 2}{\lambda} = \frac{0.693\cdots}{\lambda}$$ [27-9]

It is clear from [27-7] that $T_{1/2}$ also represents the time after which a radioactive sample has lost half its particles by decay. Unstable particles show an enormous range in half-lives. They run from less than 10^{-20} s for some unstable elementary particles to more than 10^{10} years for certain unstable heavy nuclei found in nature and facing decay since the formation of the universe some 10 billion years ago.

If a particle has not decayed during one half-life, what is the probability that it will survive for another time period of $0.693/\lambda$ (another half-life)? The chances are again fifty-fifty. This implies that at any given time one unstable particle is like any other of the same species, quite apart from how long a period of time each has already lived. After all, particles, unlike human beings, have no way of knowing how long they have survived. For particles, each new moment of time is a new beginning. This simple assertion, in fact, is a basis for *deducing* the law of radioactive decay, as is done in Example 27-1. Although the decay constant and the related half-life tell us the probability for survival of particles of a given species, there is absolutely no way of predicting when any one particle will decay. Indeed, because $e^{-\lambda t}$ becomes exactly zero only after an infinite time has elapsed, it is possible, although not likely, that a particle with a very short half-life may actually survive decay over very long times.

Mean life defined

The half-life is the time interval for which the chances for decay and survival are equal, but is not the average lifetime, usually called the *mean life*. It is easy to see that the mean life of a particular species must, in fact, *exceed* the half-life: since some few particles can survive decay over intervals approaching infinite times, the average lifetime of a group of particles is longer than the time after which half of the initial particles will have died. A more detailed consideration (Problem 27-2) shows that the mean life τ is precisely the reciprocal of the decay constant λ:

[27-10] $$\tau = \frac{1}{\lambda} = \frac{T_{1/2}}{0.693} = 1.44\, T_{1/2}$$

Thus, the mean life exceeds the half-life by 44 percent (see Fig. 27-13).

A completely stable particle is characterized by an infinite mean life. For an unstable particle the mean life gives a quantitative measure of its staying power against decay in time; e.g., a free neutron, one outside a nucleus, has a mean life of 15.5 min. We shall find in the next chapter that the mean life of any of the elementary particles in nature is as fundamental a property as its rest mass or electric charge.

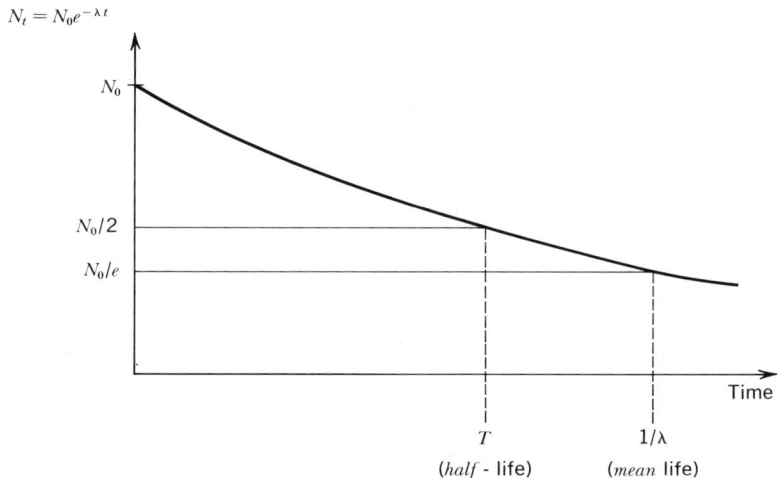

FIGURE 27-13 The exponential decay with time of the number N_t of unstable particles. The half-life T is that time interval after which $N_t = \frac{1}{2}N_0$; the mean life $1/\lambda$ is that time interval after which $N_t = N_0/e$.

c **EXAMPLE 27-1**

(*a*) In the text we have seen that the observed exponential decay in the activity of a radioactive material leads to the conclusion that for an unstable particle one moment in time is like any other. Now show the reverse: derive the mathematical form of the law of radioactive decay from the assumptions that all unstable particles of a collection of one species are identical and indistinguishable and that the chance for decay or survival of any particle is not dependent on its past history—each new moment of time is a new beginning. (*b*) Show that the activity A_t is related to the number of surviving unstable particles N_t at that instant and to the decay constant λ by the relation $A_t = \lambda N_t$.

SOLUTION

a   The total number of unstable nuclei N is large; so is the number dN decaying over a short time interval dt. The fractional number of particles decaying per unit time is then $(dN/N)/dt$. Since the decay is assumed not to depend on any past history of survival, the fractional number of particles decaying per unit time must be the *same* for *all* instants of time; i.e., it must be a *constant*. Therefore,

$$\frac{dN/N}{dt} = -\lambda$$

where λ is a constant. The minus sign appears since N is *decreasing* because of decay.

We can rewrite the above expression as

$$\frac{dN}{N} = -\lambda\, dt$$

and then integrate it, choosing limits on the left side for the number of surviving particles from an initial N_0 to a final N_t and on the right side from an initial time 0 to the final time t.

$$\int_{N_0}^{N_t} \frac{dN}{N} = -\int_0^t \lambda\, dt$$

$$\ln \frac{N_t}{N_0} = -\lambda t$$

$$N_t = N_0 e^{-\lambda t}$$

Radioactive-decay law derived

which is the decay law [27-7].

b Taking the time derivative of the equation above, we have

$$\frac{dN_t}{dt} = -\lambda N_0 e^{-\lambda t} = -\lambda N_t$$

But dN_t/dt is the change in the number of *surviving* particles per unit time: therefore, $-dN_t/dt$ is the number of particles *emitted* per unit time, or the activity A_t. The above equation then becomes

$$A_t = \lambda N_t = \lambda N_0 e^{-\lambda t} = A_0 e^{-\lambda t}$$

where, as before, $A_0 = \lambda N_0$.

+ 27-7 NUCLEAR REACTIONS

In nuclear reactions, as opposed to ordinary chemical reactions, the participating particles come so close together that the dominant force between them is the *nuclear* force. The nucleons can form new configurations, so that different structures come out than went in. For example, in the Chadwick experiment alpha particles, or ^4_2He nuclei, were directed at a target of beryllium, and neutrons emerged. We may think of an alpha particle as coming close enough to the nucleus of a beryllium atom for the two particles to amalgamate under the action of the nuclear force. The resultant collection of nucleons is not stable, however, and it decays into a more stable configuration by ejecting a neutron, leaving a nucleus of carbon 12. The overall reaction can be written

$$^{9}_{4}\text{Be} + ^{4}_{2}\text{He} \rightarrow ^{1}_{0}n + ^{12}_{6}\text{C}$$

As in the decay reactions previously considered, the balancing of subscripts and of superscripts indicates that electric charge and total number of nucleons, respectively, are conserved.

Every nuclear reaction must be consistent with energy conservation. We have already seen that spontaneous reactions which start with particles at rest are energetically possible only if the initial rest mass (or rest energy) is greater than or equal to the final rest mass (or rest energy). For more general nuclear reactions the criterion is that the total energy into the reaction (the rest energies of the particles into the reaction together with their kinetic energies) must at least equal the total rest energy of the particles that emerge from the reaction. Thus the photodisintegration of the deuteron

$$\gamma + ^{2}_{1}\text{H} \rightarrow ^{1}_{1}\text{H} + ^{1}_{0}n$$

which is the inverse of [27-4], cannot occur unless the gamma-ray photon brings in at least 2.2 MeV of (kinetic) energy.

Within the framework of total energy conservation it is possible to have kinetic energy created (at the expense of rest energy) or destroyed (to the gain of rest energy) in a nuclear reaction. *Nuclear fission* and *nuclear fusion* are two types of reactions which generate kinetic energy. The shape of the binding-energy curve—elements of middle weight having a binding energy per particle greater than either light or heavy elements (Fig. 27-4)—already indicates that both the combining (fusion) of lightweight elements into those of middle weight and the splitting (fission) of heavy elements into those of middle weight would mean a gain in binding energy, hence a loss of rest energy. Therefore, kinetic energy could be produced in the reaction.

Consider the fission of the very heavy nucleus, uranium 235. If a $^{235}_{92}\text{U}$ nucleus has a neutron directed at it, the neutron is attracted by the nuclear force when sufficiently close and the capture of the neutron leads to the formation of a highly unstable nucleus, $^{236}_{92}\text{U}$. Indeed, its instability is so great that it spontaneously splits into two (or more) nuclei of smaller size and at the same time releases a few neutrons. A typical form of the reaction might be

$$^{235}_{92}\text{U} + ^{1}_{0}n \rightarrow ^{236}_{92}\text{U} \rightarrow ^{94}_{36}\text{Kr} + ^{140}_{56}\text{Ba} + 2\,^{1}_{0}n$$

with nuclei of barium and krypton formed, together with two additional free neutrons. Energy of 200 MeV is released in this reaction in the form of the kinetic energy of the released particles. Under proper circumstances the emitted neutrons may initiate still other fission reactions with other uranium nuclei. A chain reaction may ensue, in which the energy generated exceeds that from a typical chemical reaction by a factor of about 10^6; the sudden release of this energy is illustrated in a nuclear bomb, the slow release in a nuclear reactor. The thermal-energy output of the reactor may in turn be used to produce electric power.

Two examples of nuclear reactions in which lightweight nuclei are fused together with the release of relatively large amounts of energy are

$$^2_1H + {}^2_1H \rightarrow {}^3_2He + {}^1_0n$$

and

$$^2_1H + {}^3_1H \rightarrow {}^4_2He + {}^1_0n$$

In the first reaction two deuterons combine to yield a helium 3 nucleus and a neutron; the kinetic energy of the outgoing particles exceeds the kinetic energy of the deuterons by 3.3 MeV, as can be verified by comparing the rest masses of the four particles. In the second reaction a nucleus of deuterium (hydrogen 2) and a nucleus of tritium (hydrogen 3) combine to yield a helium 4 nucleus and a neutron; the net energy release in this reaction is 17.6 MeV.

Such reactions can take place, however, only if rather special conditions are met. The nuclei will interact through the nuclear force only if the kinetic energies of the combining particles are sufficiently high to make the particles approach each other to within about 3×10^{-15} m, despite the Coulomb repulsion that exists between the positively charged nuclei at large distances. One might, for example, use a high-energy accelerating machine to project one kind of particle against a target of the second kind, but only a very few incident particles will then actually induce nuclear reactions because of the very small area the target particles present to incident particles. Or, to produce the very high kinetic energies required, one might raise the temperature and hence the average kinetic energy per particle. Kinetic energies of around 1 MeV imply temperatures of millions of degrees Kelvin; such very high ignition temperatures are found naturally only in stars, where fusion-type nuclear reactions are responsible for starlight. These temperatures can be achieved on earth by a nuclear-fission explosion. Indeed, a hydrogen bomb contains a nuclear-fission trigger; once the high temperature is achieved, the hydrogen fuel explodes by nuclear *fusion*.

For the fusion process to take place nonexplosively, in controlled fashion, it is thus necessary that the particles somehow be confined while at stellar temperatures. All ordinary containers are unsuitable, not so much because the very hot contents would melt the container but because the cool container would chill the contents below the temperature for spontaneous nuclear fusion. One possible solution for electrically charged particles is a *magnetic bottle*, an arrangement of magnetic fields in which the particles are deflected by magnetic forces and made to stay within some limited region of space. The Van Allen belts which trap electrons near the earth (see Example 15-4) are realizations on a large scale of the magnetic bottle.

SUMMARY

Protons and neutrons, of which all nuclei consist, attract each other by a very strong nuclear force but only when separated by distances less than about 10^{-15} m. The binding of nucleons within nuclei is so strong that the energy of binding, typically of the order of several MeV, can be computed directly, by applying mass-energy conservation, from a comparison of rest masses.

The radioactive decay processes of alpha, beta, and gamma emission involve the spontaneous emission, respectively, of a helium nucleus, an electron (or positron) together with an antineutrino (or neutrino), and a photon. The neutrino (or antineutrino) has zero electric charge and rest mass; it travels at speed c and interacts only feebly with other particles. Any decay process is possible only if the laws of energy, momentum, and charge conservation are satisfied.

The probability for the decay of unstable particles follows the law of radioactive decay according to which the activity of a radioactive sample or the number of surviving unstable particles decays exponentially with time, the same fractional decrease occurring over all equal time intervals. The probability that a single unstable particle will survive for at least t more seconds is $e^{-\lambda t}$ (see [27-8]), where the decay constant λ controls the rate of decay. The decay of an unstable particle is independent of its past history, and its mean life $1/\lambda$ is one of its fundamental properties.

In a nuclear reaction the participating nucleons form new configurations so that the nuclei emerging from a nuclear reaction may differ from those entering. Kinetic energy may be released or consumed in a nuclear reaction, depending upon whether the total rest energy of the particles decreases or increases. In nuclear fission very heavy nuclei are split apart, and in nuclear fusion light nuclei are amalgamated, both with the release of kinetic energy of the order of an MeV per nucleon.

CHAPTER 27
NUCLEAR PHYSICS

PROBLEMS

27-1 Show, by comparing the atomic masses of ^{12}C (12.000000 u) and ^{12}B (12.014354 u), that the energy released in the decay of a ^{12}B nucleus is 13.37 MeV.

c 27-2 Show that the mean life of a species of radioactive nuclei having a decay constant λ is $1/\lambda$. *Hint:* the mean life τ is defined as

$$\tau = \int_{N_0}^{0} t \, dN \Big/ \int_{N_0}^{0} dN$$

27-3 Verify (*a*) that the kinetic energy released in the nuclear reaction ^{2}H + ^{2}H → ^{3}He + ^{1}n is 3.3 MeV and (*b*) that the energy released in the reaction ^{2}H + ^{3}H → ^{4}He + ^{1}n is 17.6 MeV. See Table 27-4 for atomic masses.

27-4 (*a*) Use the uncertainty principle to show that an electron confined within a spatial extent not exceeding 10^{-15} m has an energy of at least 10^2 MeV. (*b*) What is the minimum energy of a proton or neutron confined within the same distance?

27-5 It is observed that when photons having an energy of 3.73 MeV irradiate deuteron nuclei, protons emerge with a kinetic energy of 0.75 MeV. Since the neutrons have nearly the same mass as the protons, the neutrons produced in the photodisintegration of deuterons by such photons also have a kinetic energy of 0.75 MeV. Use these data, together with accurately known masses of the proton and deuteron, to *compute* the mass of the neutron. This illustrates the procedure that is used to arrive at highly precise values for the neutron mass.

27-6 As Fig. 27-4 shows, the average binding energy per nucleon for nuclei of uranium is 7.5 MeV, while the comparable figure for nuclei of half the mass is about 8.5 MeV. Use these data to show that the energy released in a typical fission of a uranium nucleus is of the order of 200 MeV.

27-7 In the first identified nuclear reaction (in 1919 by Lord Rutherford) nitrogen 14 nuclei were bombarded by alpha particles, and protons were observed to emerge. What was the massive nucleus produced in each such reaction?

27-8 The first nuclear reaction induced by particles from a high-energy accelerator was observed by J. D. Cockcroft and E. T. S. Walton in 1932. They fired protons with a kinetic energy of 0.50 MeV at lithium 7 nuclei and observed alpha particles emerge from the reaction. Taking the lithium nuclei to be free and initially at rest, what was the total kinetic energy of the two alpha particles produced in each reaction? See Table 27-5.

27-9 Carbon 14 is radioactive and decays by beta emission with a half-life of 5,740 years. What fraction of an initial collection of ^{14}C atoms will have survived after a period of 22,960 years?

Particle	Atomic Mass, u
^1n	1.008665
^{2}H	2.014102
^{3}H	3.016050
^{3}He	3.016030
^{4}He	4.002603

TABLE 27-4

Particle	Atomic Mass, u
^{1}H	1.007825
^{4}He	4.002603
^{7}Li	7.016004

TABLE 27-5

Particle	Atomic Mass, u
^{1}H	1.007825
^1n	1.008665
^{11}C	11.011432
^{11}B	11.009305
^{12}C	12.000000

TABLE 27-6

27-10 It is shown in the text that the average binding energy of the 12 nucleons in the nucleus of ^{12}C is 7.68 MeV. (*a*) Suppose that just one neutron is removed from ^{12}C to leave the nucleus ^{11}C. How much work is required, i.e., what is the binding energy of this one neutron? (*b*) Suppose now that a proton is removed from a ^{12}C nucleus, leaving a nucleus of ^{11}B. How much work is required, i.e., what is the binding energy of this one proton? See Table 27-6 for atomic masses.

27-11 The probability that a given unstable particle will survive one more half-life is 1/2. What is the probability that a given unstable particle will survive a time interval equal to its mean life.

27-12 The ratio of radioactive carbon (^{14}C) to stable carbon is fairly constant in *living* matter, but in *dead* matter the ^{14}C nuclei decay with the half-life 5,740 years. Explain how radioactivity measurements are useful in dating archeological remains of matter that was once living.

ELEMENTARY-PARTICLE PHYSICS

CHAPTER TWENTY-EIGHT

Behind whatever happens, seen or unseen, the physicist finds the interaction of particles. This leads him to hope that fundamental physics will be "solved" once all the players (the various elementary particles) have been identified and understood, and the rules of the game (the conservation laws) have been described. Whether or not such ultimate insight can ever be reached, the search for it stands at the forefront of contemporary physics.

Elementary-particle physics is remarkable in several ways. First, the number of particles now regarded as fundamental is surprisingly large—perhaps too large, if one is to think of the universe as ordered and simple. Familiar species, e.g., the electron and the proton, are stable, but many recently discovered particles have a transitory existence, lasting for less than 10^{-23} s. While these particles themselves are submicroscopic and elusive, the apparatus with which they are produced and studied is usually enormous, both in size (miles) and in cost (hundreds of millions of dollars; see Fig. 28-1).

Although the chemical elements and their properties can now be systematically accounted for on the basis of the electric interaction and the rules of the quantum theory, no master plan has yet been perceived which takes in all the elementary particles. Nevertheless, the field of elementary-particle physics involves some of the most sophisticated techniques of both experimental and theoretical physics. In the present chapter, therefore, we merely illustrate a few of the simpler and most basic aspects of particle physics.

The most familiar of the elementary particles, whose properties are now well established and which can be organized into several families, are shown in Table 28-1. This chapter is devoted to showing that the entries in this table make sense, that there is some degree of coherence to be found in the properties and interrelationships of elementary particles.

28-1 THE UNIVERSAL CONSERVATION LAWS

The conservation laws of physics have been the recurring theme of this book. Their importance lies in their simplicity and generality: in every instance, some physical property stays the same for an isolated system, regardless of what goes on inside the system. We speak, for example, of the *law* of linear-momentum conservation because, up to this time, no exception to it has ever been found, and it is a safe bet that momentum conservation will apply in still one more case. Of course, if one clear violation of a supposed law of physics is found, this "law" will be rendered invalid.

SECTION 28-1
THE UNIVERSAL
CONSERVATION
LAWS

FIGURE 28-1 (*a*) Exterior and (*b*) interior views of the Stanford electron linear accelerator. This machine, 2 miles in length, and costing about $100 million, accelerates electrons to 20 GeV, at which energy the electron speed is less than that of light by only 3 parts in 10^{10}. (*Stanford Linear Accelerator Center, Stanford University.*)

(*a*)

(*b*)

CHAPTER 28 ELEMENTARY-PARTICLE PHYSICS

Name	Particle	Electric Charge e	Mass m_e	Spin Angular Momentum $h/2\pi$	Family number Electron	Family number Muon	Family number Baryon	Mean Life s	Antiparticle
Baryon family									
Hyperons:									
Omega	Ω	−1	3,272	$\frac{3}{2}$	0	0	1	1.3×10^{-10}	$\overline{\Omega}$
Xi	Ξ	−1	2,585	$\frac{1}{2}$	0	0	1	1.7×10^{-10}	$\overline{\Xi}$
		0	2,573	$\frac{1}{2}$	0	0	1	3.0×10^{-10}	
Sigma	Σ	−1	2,342	$\frac{1}{2}$	0	0	1	1.6×10^{-10}	$\overline{\Sigma}$
		0	2,333	$\frac{1}{2}$	0	0	1	$<1.0 \times 10^{-14}$	
		1	2,327	$\frac{1}{2}$	0	0	1	0.8×10^{-10}	
Lambda	Λ	0	2,184	$\frac{1}{2}$	0	0	1	2.5×10^{-10}	$\overline{\Lambda}$
Nucleons:									
Neutron	n	0	1,839	$\frac{1}{2}$	0	0	1	9.3×10^{2}	$\overline{n}$
Proton	p	1	1,836	$\frac{1}{2}$	0	0	1	Stable ($>2 \times 10^{28}$ years)	$\overline{p}$
Mesons									
Kaon	K	0	974	0				0.8×10^{-10}; 5×10^{-8}	$\overline{K}$
		1	966	0				1.2×10^{-8}	
Pion	π	1	273	0				2.6×10^{-8}	$\overline{\pi}$
		0	264	0				0.9×10^{-16}	
Muon family									
Muon	μ	−1	207	$\frac{1}{2}$	0	1	0	2.2×10^{-6}	$\overline{\mu}$
Muon neutrino	ν_μ	0	0	$\frac{1}{2}$	0	1	0	Stable	$\overline{\nu}_\mu$
Electron family									
Electron	e	−1	1	$\frac{1}{2}$	1	0	0	Stable ($>2 \times 10^{21}$ years)	$\overline{e}$
Electron neutrino	ν_e	0	0	$\frac{1}{2}$	1	0	0	Stable	$\overline{\nu}_e$
Photon	γ	0	0	1				Stable	Itself
Graviton (?)	g	0	0	2				Stable	Itself

TABLE 28-1

The conservation laws now thought to be universally valid, i.e., applicable for all particles and under all circumstances, are the following:

1. (Linear) Momentum
2. Mass-energy
3. Electric charge
4. Angular momentum
5. Electron number
6. Muon number
7. Baryon number

The first four laws are familiar; the last three will be developed in this chapter. In addition to these universally valid conservation laws, other conservation laws have been discovered which apply to some but not all interactions between elementary particles. One example, the law of parity conservation, will be discussed in Section 28-8.

The conservation laws are basically rules of prohibition. When two billiard balls collide, we know that the total momentum and energy into the collision must equal the total momentum and energy out of the collision. Collisions that would violate momentum or energy conservation (or electric-charge or angular-momentum conservation) are ruled out. Nevertheless, there is a whole variety of speeds and directions with which the balls may emerge from a collision without violating the conservation laws. As we shall see, the view of reality that emerges in the study of the elementary particles is this: unless some conservation law precludes a process, that process will occur; in other words, *all possible* processes will occur. This is not to say that the probability of occurrence is the same for all allowed events. In general, when we inquire whether a certain conceivable process will take place, we are, in effect, asking whether all the conservation laws are satisfied. If experiment shows that some process in fact, never occurs, we have the making of still another heretofore undiscovered conservation law.

Unless ruled out by a conservation law, all conceivable processes take place.

28-2 PAIR PRODUCTION AND PAIR ANNIHILATION

A photon cannot spontaneously turn into a single charged particle; that would violate the law of charge conservation. How about the process in which a photon goes out of existence and simultaneously two particles with equal but opposite charges are created? This is not prohibited by the conservation laws—indeed, it happens. In the process of *pair production*, a photon disappears while an electron and a positron, a particle of the same mass as the electron but with an electric charge $+e$, are produced. The positron, which is identical with the electron in all respects but the sign of its charge (and properties related to the sign of its charge), is said to be the *antiparticle* of the electron. This terminology relates more closely to the inverse process, *pair annihilation*, in which an electron and a positron destroy each other, with the creation of photons. As we shall see, still other particles have their antiparticles, identical in mass but opposite in charge. Clearly, the particle-antiparticle relation is symmetric: if B is the antiparticle of A, then A is the antiparticle of B.

Pair production: photon into an electron-positron; pair annihilation: electron-positron into photons

The rest energy of an electron (or positron) is 0.51 MeV. It follows that a photon can create an electron-positron pair only if its energy $h\nu$ equals or exceeds $2(0.51 \text{ MeV}) = 1.02 \text{ MeV}$ (mass-energy conservation).

CHAPTER 28
ELEMENTARY-PARTICLE PHYSICS

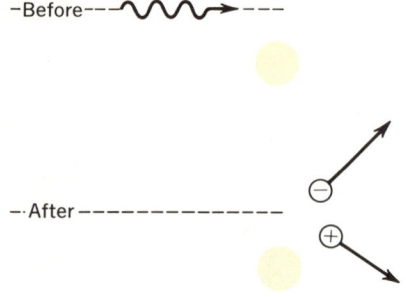

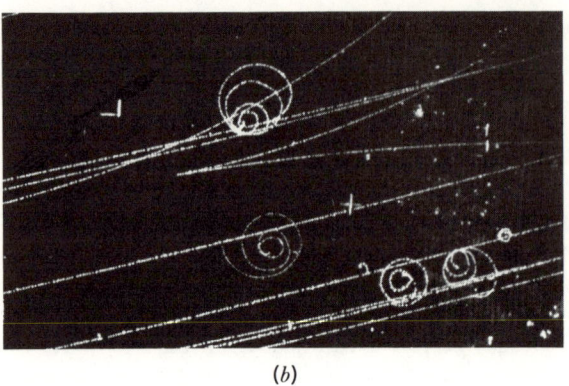

FIGURE 28-2 (*a*) A photon annihilates into an electron-positron pair. (*b*) Photograph showing tracks by electron-positron pairs moving in a magnetic field; note that the positive and negative particles are deflected in opposite directions (look at Figure 15-18*d*). (*University of California, Lawrence Radiation Laboratory, Berkeley.*)

We may write

$$\gamma \longrightarrow e + \bar{e} \qquad [28\text{-}1a]$$

Electric charge in units of e: $\quad 0 = (-1) + (+1)$

Mass-energy: $\quad 1.02 \text{ MeV}_{\min} = 0.51 \text{ MeV}_{\min} + 0.51 \text{ MeV}_{\min}$

where we have represented the photon by γ. The positron, or electron antiparticle, is represented by $\bar{e}$; putting a bar over the symbol for a particle is the usual way to represent its antiparticle.

Can an *isolated* photon simply turn into an electron-positron pair? It is not difficult to show (see Example 28-1) that it cannot, inasmuch as such an event would violate the momentum-conservation law. Pair production occurs only when the photon comes close to some massive object, such as a nucleus (see Fig. 28-2). The nucleus acquires enough momentum to assure momentum conservation but because of its large mass does not gain sufficient kinetic energy to disturb the energy balance between photon, electron, and positron.

EXAMPLE 28-1 Can a photon in *empty space* spontaneously change into an electron-positron pair? Or must a massive particle also be present so that the momentum conservation law will not be violated?

SOLUTION Suppose, for the sake of demonstration, that an isolated photon with momentum p_γ and energy $E_\gamma = p_\gamma c$ spontaneously produces an electron and positron, each with total energy E_e and each traveling in the same direction as the incident photon. Then energy conservation requires that

$$E_\gamma = 2E_e$$

or

$$p_\gamma c = 2E_e \qquad [28\text{-}2]$$

and momentum conservation requires that

$$p_\gamma = \frac{2\sqrt{E_e^2 - E_0^2}}{c} \qquad [28\text{-}3]$$

where we recognize that the relativistic momentum of an electron or positron with rest mass E_0 is given by $\sqrt{E_e^2 - E_0^2}/c$. If [28-2] and [28-3] are both to be satisfied, then

$$\underbrace{\frac{2E_e}{c}}_{p_\gamma \text{ from } [28\text{-}2]} = \underbrace{\frac{2\sqrt{E_e^2 - E_0^2}}{c}}_{p_\gamma \text{ from } [28\text{-}3]} \qquad [28\text{-}4]$$

which implies that $E_0 = 0$, in contradiction to the fact that the electron and positron have a finite rest mass. Thus, the hypothetical process cannot take place.

If, however, the incident photon is close to a massive particle, such as a nucleus, when the electron-positron pair is created, some of the photon's momentum and energy can be transmitted to the massive particle. If the massive particle, initially at rest, acquires momentum in the forward direction, [28-3] must be rewritten with an additional term appearing on the right-hand side

$$p_\gamma = \frac{2\sqrt{E_e^2 - E_0^2}}{c} + p_M \qquad [28\text{-}3a]$$

And at the same time [28-2] must also be rewritten with an additional term representing the change in energy ΔE_M of the massive particle

$$E_\gamma = 2E_e + \Delta E_M \qquad [28\text{-}2a]$$

It is not difficult to show that the new term ΔE_M is negligible compared to the energy E_γ of the photon and the energy $2E_e$ of the electron-positron pair. As a result [28-2] is virtually unchanged, and [28-4] becomes

$$\underbrace{\frac{2E_e}{c}}_{p_\gamma \text{ from } [28\text{-}2]} = \underbrace{\frac{2\sqrt{E_e^2 - E_0^2}}{c} + p_M}_{p_\gamma \text{ from } [28\text{-}3a]} \qquad [28\text{-}4a]$$

This equation, in contrast to [28-4], *can* be satisfied using the known nonzero rest energy E_0 of an electron: the extra momentum p_M acquired by the massive particle increases the right-hand side by the amount necessary to provide the required total value $2E_e/c$.

Finally, to verify that the new term ΔE_M is indeed negligible compared to the other energies in [28-2a], note that the momentum p_M acquired by the massive particle must be less than the available momentum of the incident photon. Moreover, in a typical situation the energy of the incident photon is a few MeV; for example, 2.0 MeV, or somewhat in excess of the 1.02 MeV minimum energy required for the pair production. Thus we can write the change in energy of the massive particle, which is just the nonrelativistic kinetic energy it acquires, as

[28-5] $$\Delta E_M = \frac{1}{2} M_0 V^2 = \frac{1}{2} M_0 \left(\frac{p_M}{M_0}\right)^2 = \frac{(p_M c)^2}{2 M_0 c^2}$$

But, from what has just been said, $p_M c < p_\gamma c = E_\gamma = 2.0$ MeV. Moreover, the rest energy $M_0 c^2$ of even the smallest atom (hydrogen) is nearly 1,000 MeV. Therefore,

$$\Delta E_M < \frac{2.0^2 (\text{MeV})^2}{2(1,000 \text{ MeV})} = 0.002 \text{ MeV}$$

which is indeed negligible compared to the 2.0 MeV photon energy or the 1.0 MeV energy of either member of the pair.

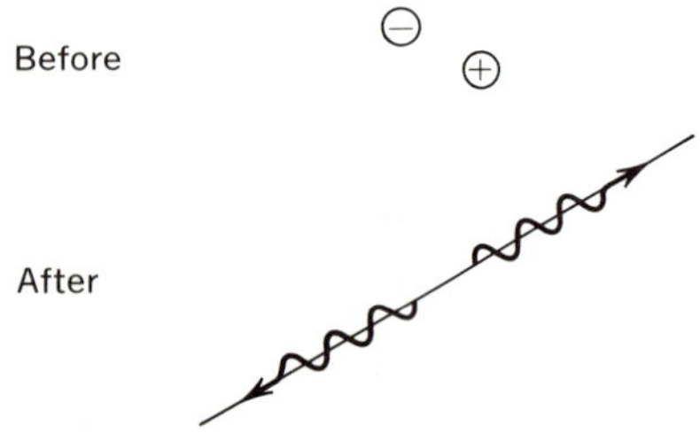

Positrons are vastly outnumbered by electrons, at least in our corner of the universe. The imbalance is maintained by the process of pair annihilation. Once created, a positron in isolation is just as stable as an electron. However, a typical positron loses its kinetic energy in collision with other particles, comes to rest, and annihilates with a stationary electron (Fig. 28-3). Not one, but at least two, photons are created as a result (a single photon has positive momentum, whereas the momentum before the annihilation is zero). Usually, two photons appear, with equal and opposite momenta and therefore equal energies. We write

[28-1b] $e + \bar{e} \rightarrow \gamma + \gamma$

Here the value of the photon energies is 0.51 MeV, which is the rest energy of an electron or positron, in accordance with energy conservation.

FIGURE 28-3 An electron-positron pair annihilates into two photons.

Besides pair production and pair annihilation we are already familiar with two other types of reaction involving photons and electrons (or positrons). The first of these is basically the atomic photoelectric effect (Section 23-3); e.g., a hydrogen atom absorbs a photon and is broken into a separated proton and electron

$$h\nu + {}_1^1\text{H} \to p + e$$

If the photon energy is much greater than the binding energy of the atom, we can associate the energy change primarily with the electron and write

$$e + \gamma \to e' \qquad [28\text{-}1c]$$

where the electron on the right has a different energy from that on the left (Fig. 28-4a).

Actually, a *free* electron cannot absorb a photon because such a process would not simultaneously satisfy the conservation of energy and of momentum (the reasoning is similar to that in Example 28-1). Nevertheless, we are going to grant that the process is conceivable! The basis for this mad-sounding assertion is to be found in the uncertainty principle, which in fact allows fleeting evasions—although no standing violation—of the conservation laws. We shall discuss this point in some detail in Section 28-7; for now, the following consideration may help. Suppose we set out to detect an "illegal" event, one in which an amount of energy ΔE is created from nothing. To do so, we would certainly have to be able to achieve a precision in energy measurements of ΔE or better. But according to [25-12], $\Delta E\, \Delta t \gtrsim h$, the measurement would have to extend over a time period of at least $\Delta t \simeq h/\Delta E$. If *before that period had elapsed* a second event took place, in which an amount of energy ΔE was *destroyed*, no net creation of energy would be recorded and the violation would forever remain below the threshold of observability.

A second type of electron-photon reaction is exemplified by atomic radiation: an electron, in the presence of a massive object, loses energy, thereby creating a photon (Fig. 28-4b). In symbols,

$$e \to e' + \gamma \qquad [28\text{-}1d]$$

Again, the electron energies are different on the two sides of the equation.

We can unite the basic electron-photon interactions through the concept of the *electron family*. The members of the electron family with

28-3 CONSERVATION OF ELECTRON NUMBER

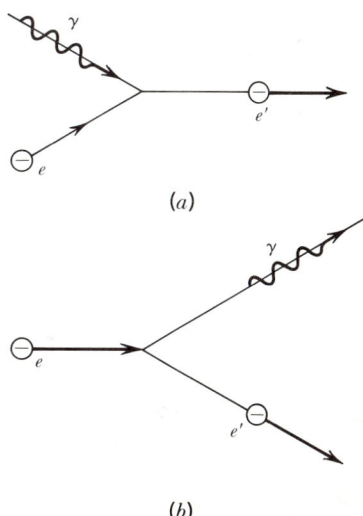

FIGURE 28-4 (a) An electron e absorbs a photon γ and thereby becomes an electron e' with different energy and momentum; (b) an electron e emits a photon γ and becomes a different electron e'.

Electron number: for an electron $+1$, for a positron -1

CHAPTER 28 ELEMENTARY-PARTICLE PHYSICS

TABLE 28-2

Reaction	Equation Number	Equation	Conditions
Pair production	[28-1a]	$\gamma \to e + \bar{e}$	Near a massive nucleus
Electron number		$0 = (+1) + (-1)$	
Pair annihilation	[28-1b]	$e + \bar{e} \to \gamma$	Two or more γ
Electron number		$(+1) + (-1) = 0$	
Photon absorption	[28-1c]	$e + \gamma \to e$	Near a massive nucleus
Electron number		$(+1) + 0 = (+1)$	
or		$\bar{e} + \gamma \to \bar{e}$	Near a massive nucleus
Electron number		$(-1) + 0 = (-1)$	
Photon production	[28-1d]	$e \to e + \gamma$	Near a massive nucleus
Electron number		$(+1) = (+1) + 0$	
or		$\bar{e} \to \bar{e} + \gamma$	Near a massive nucleus
Electron number		$(-1) = (-1) + 0$	

which we are concerned here are the electron, which is assigned an electron number of $+1$, and the antielectron (positron), which is assigned an electron number of -1. The photon is not a member of the electron family, so that the electron number for it is zero.

Writing down the processes once more and adding the electron number in each instance, we have the results shown in Table 28-2. It is seen that the *total* number is preserved in each equation. Evidently a conservation law is operating here.

We note, moreover, that (1) if a particle appearing on one side of a reaction equation is shifted to the other side and (2) at the same time it is converted to the antiparticle (or the antiparticle to the particle), not only is electron-number conservation still applicable but the new equation represents another of the basic processes. For example, starting with the pair-production reaction

$$\gamma \longrightarrow e + \bar{e}$$
$$0 = (+1) + (-1)$$

and then shifting the positron $\bar{e}$ to the left side, we obtain

$$e + \gamma \longrightarrow e$$
$$(+1) + 0 = (+1)$$

which is the photon-absorption relation.

Indeed, examination of the reaction relations shows that all of them can be obtained from any one of them through this procedure if, in addition, the direction of the arrow can be reversed. Reversing the arrow implies that the process is imagined to be run backward in time. Thus, pair annihilation visualized as proceeding as one would see it in a movie of the events run the "wrong" way is pair production; in one case the particle and antiparticle create a photon, and in the other the photon creates the particle-antiparticle pair.

The reversibility in time of all processes at the submicroscopic level has already been discussed in our treatment of the kinetic theory of gases. There we saw that if one were given a motion-picture film of events taking place between interacting molecules, there would be absolutely no way of telling which of the two possible ways of running the film through the projector is "correct." Both are equally in conformity with the laws of physics. What time reversibility, together with the particle-antiparticle shift rule, implies for the various electron-photon processes is this: they are all basically the *same* process. Depending on whether this master process is conceived to go forward or backward in time and upon whether a given event is described as the appearance of a particle or as the disappearance of an antiparticle, one or another of the four reactions is observed.

Time reversibility of submicroscopic processes

All electron-photon processes illustrate one basic process.

So far, nothing at all has been said which might distinguish the electron number from the electric charge (measured in units of $-e$). Yet we have claimed that conservation of electron number represents a new and independent law of nature. To see what kind of support this assertion can be given, consider the following hypothetical process, the decay of a proton into a positron and a photon:

$$p \rightarrow \bar{e} + \gamma$$

Although in actual fact this decay process is never found to occur, it is not in violation of the conservation laws of linear momentum (products flying off in opposite directions), energy (the proton rest energy exceeds that of the positron), or electric charge (both proton and positron have charge $+e$). It can also be shown that the process is consistent with the fourth conservation law, that of angular momentum.

If the basic attitude to be taken in considering the behavior of elementary particles (and therefore all of physics) is that any process which is not forbidden by these four conservation laws *must* occur, then the decay of a proton into a positron and photon should certainly be expected to occur. Yet the results of observation are negative: a proton never decays, and if it does not, there must be some rule, some conservation law, which prohibits it. The electron-number conservation law will be just such a rule if the proton is assigned an *electron* number of *zero*; that is, it does not belong to the family. Thus, it is the fact that the proton does not decay (and that certain other conceivable processes do not take place) which leads to the idea of an electron number, the number to be assigned to various particles, and the conservation law itself.

28-4 CONSERVATION OF BARYON NUMBER

Baryons: particles with proton mass or greater

The most massive particles listed in Table 28-1 are known as *baryons*. In addition to the familiar proton and neutron, there are several unstable and more massive members of the family—the omega, xi, sigma, and lambda particles. For each particle there is also a corresponding antiparticle.

A study of particle decays and interactions shows that just as conservation of electron number accounts for the net number of electrons and positrons which may emerge from a reaction, so too a conservation law controls the net number of baryons. If a baryon particle is arbitrarily assigned the baryon number $+1$, a baryon antiparticle the number -1, and all other particles are assigned the baryon number 0, then the total baryon number going into any reaction equals the total baryon number out of the reaction. For example, the lambda particle Λ^0 is found to decay, with a mean life of 2.5×10^{-10} s, into a proton or neutron together with a pion (pi meson) as shown in Fig. 28-5. The pion, which like the photon can be created without a limitation on the number of particles, carries a baryon number zero. Thus the lambda decay is given by

[28-6] $$\Lambda^0 \longrightarrow p + \pi^-$$

Baryon number: $(+1) = (+1) + 0$

where the superscripts give the particle's electric charge in units of $+e$ and π represents the pion.

Or

[28-7] $$\Lambda^0 \longrightarrow n + \pi^0$$

Baryon number: $(+1) = (+1) + 0$

Similarly, in the decay of the sigma particle,

SECTION 28-4
CONSERVATION OF
BARYON NUMBER

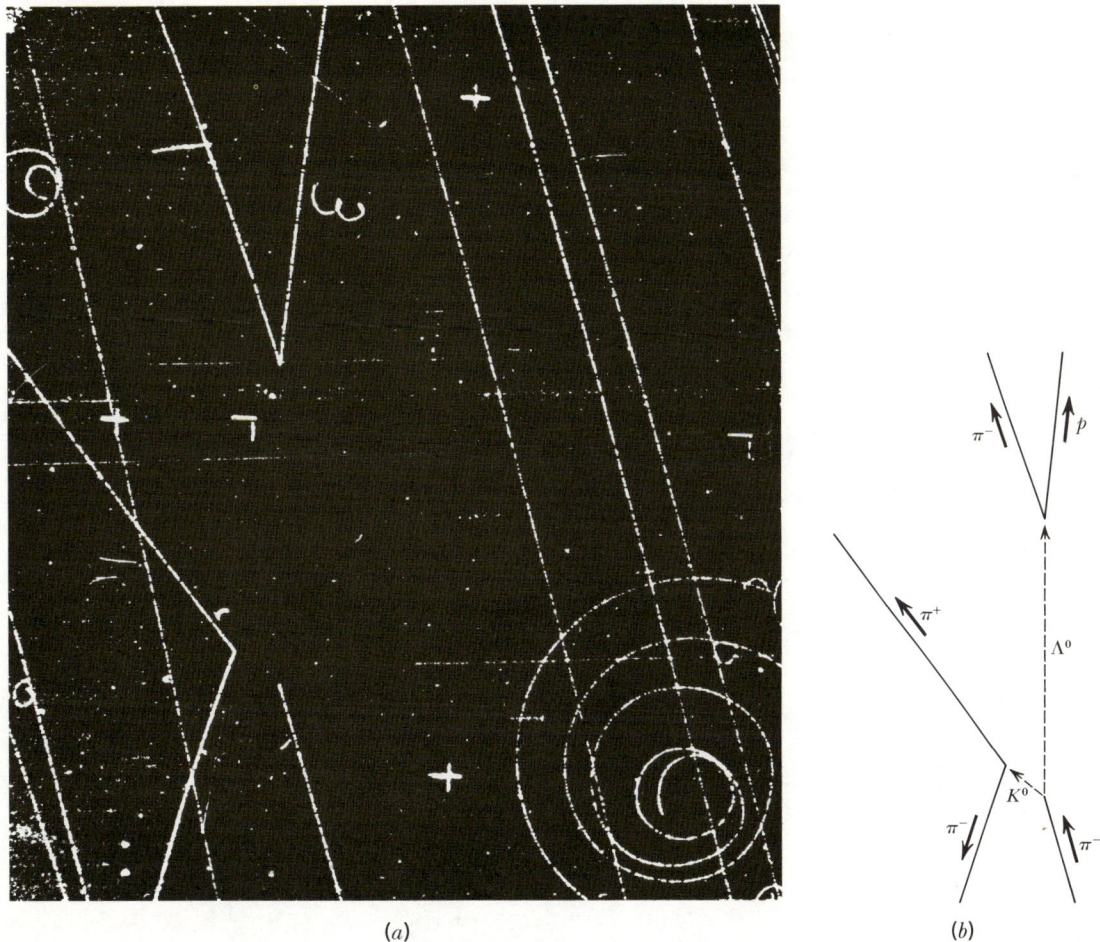

FIGURE 28-5 A bubble-chamber photograph showing the decay of a lambda particle into a proton and pion. (*University of California, Lawrence Radiation Laboratory, Berkeley.*)

$$\Sigma^0 \longrightarrow \Lambda^0 + \gamma \qquad [28\text{-}8]$$

Baryon number: $(+1) = (+1) + 0$

The reaction in which the antiproton was first produced in the laboratory involved the bombardment of a target of protons by very energetic protons (of about 6 GeV; see Example 28-2). The experiment was carried out in 1955 by Chamberlain, Segre, Wiegand, and Ypsilantis at the University of California, Berkeley. In addition to the two protons entering the reaction, a proton and an antiproton are created

[28-9]
$$p + p \longrightarrow p + p + p + \bar{p}$$
Baryon number: $(+1) + (+1) = (+1) + (+1) + (+1) + (-1)$

The antiproton was identified by noting the direction of its deflection in a magnetic field (that indicating a negatively charged particle) and by evaluating its mass indirectly through simultaneous measurements of its momentum (by the amount of deflection in a magnetic field) and its velocity (by timing the flight over a known distance).

It is also possible for a proton-antiproton pair to be created from a photon, but this reaction has a probability of occurrence even lower than the already low probability for electron-positron pair production:

[28-10]
$$\gamma \longrightarrow p + \bar{p} \quad \text{near a massive particle}$$
Baryon number: $0 = (+1) + (-1)$

Inversely, an antiproton can annihilate together with a proton to produce photons in the same fashion as an electron and a positron. Much more likely, however, is the process in which the proton and antiproton annihilate mutually to produce pions, which appear in three varieties, with electric charges $+e$, 0, and $-e$. Apart from the restrictions imposed by the energy- and charge-conservation laws, the number of pions emerging from proton-antiproton annihilation is unlimited.

For example,

[28-11]
$$p + \bar{p} \longrightarrow \pi^- + \pi^+ + 2\pi^0$$
Baryon number: $(+1) + (-1) = \quad 0 + 0 + 0$

Another possible mode of proton-antiproton annihilation is

[28-12]
$$p + \bar{p} \longrightarrow 2\pi^- + 2\pi^+ + \pi^0$$
Baryon number: $(+1) + (-1) = \quad 0 + 0 + 0$

It is again clear that pions carry baryon family number zero.

The conservation law for baryon family number, so to speak, offers "proof" that the proton, like the positron, cannot by itself decay into a lighter particle with the same positive electric charge. Actually, it is the fantastic stability of the proton (mean life $> 2 \times 10^{28}$ years) that forces upon us the concept of baryon conservation. We might also ask, for example, why a proton could not decay into a positron together with a photon or a neutral pion

or
$$p \xrightarrow{?} \bar{e} + \gamma$$
$$p \xrightarrow{?} \bar{e} + \pi^0$$

Baryon number: $(+1) \neq 0 + 0$
Electron number: $0 \neq (+1) + 0$

The decays would violate both the conservation of electron family number and of baryon family number. Thus, the proton has an effectively infinite mean life for decay.

EXAMPLE 28-2

If a sufficiently energetic proton strikes a proton initially at rest, an antiproton can be created through reaction [28-9]

$$p + p \rightarrow p + p + p + \bar{p}$$

where, in addition to the original incident proton and target proton, we have an antiproton and proton created. Show that this reaction can take place only if the incident proton's kinetic energy equals or exceeds 6 times its rest energy; i.e., the threshold energy for antiproton production is $6(0.94 \text{ GeV}) = 5.64 \text{ GeV}$.

Threshold for the creation of antiprotons

SOLUTION

The reaction above takes place only if momentum and energy are conserved. For simplicity, we view the reaction from the point of view of an observer at rest with respect to the system's center of mass. Then, as shown in Fig. 28-6a, the incident and target protons approach each other with equal momentum in opposite directions before the collision. In the center-of-mass reference frame, the *threshold* corresponds to bringing the additional two particles into existence but with none of the four particles having kinetic energy. Therefore, after the collision we find the four particles, the original two protons and the additional proton and antiproton, *at rest*.

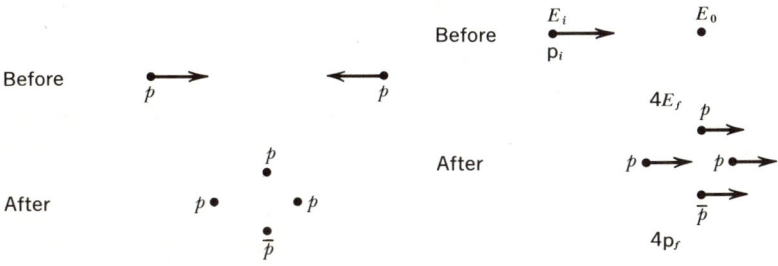

FIGURE 28-6 A proton striking a proton and producing another proton and antiproton at the threshold of the reaction $p + p \rightarrow p + p + p + \bar{p}$ as seen from the (*a*) center-of-mass and (*b*) laboratory reference frames.

This implies that in the laboratory reference frame, in which the target proton is originally at rest, all four particles appearing after the collision must move together with the *same* momentum and energy in the direction of the incident proton, as shown in Fig. 28-6b.

The initial total relativistic energy and momentum of the incident proton as observed in the laboratory we designate E_i and p_i. The rest energy of the proton, as well as that of each of the other particles, is E_0. The final total relativistic energy of any of the four particles after the reaction is denoted E_f and the common relativistic momentum p_f. Then, applying momentum and energy conservation at the threshold conditions, as observed in the laboratory, we have:

[28-13] Momentum conservation: $\mathsf{p}_i = 4\mathsf{p}_f$

[28-14] Energy conservation: $E_i + E_0 = 4E_f$

Since the relatavistic momentum is given by

[28-15] $$\mathsf{p}_i = \frac{\sqrt{E_i^2 - E_0^2}}{c}$$

and

[28-16] $$\mathsf{p}_f = \frac{\sqrt{E_f^2 - E_0^2}}{c} = \sqrt{\frac{[(E_i + E_0)/4]^2 - E_0^2}{c}}$$

we can write [28-13] as

[28-17] $$\frac{\sqrt{E_i^2 - E_0^2}}{c} = 4\sqrt{\frac{[E_i + E_0)/4]^2 - E_0^2}{c}}$$

After a little algebra we find that

[28-18] $E_i = 7E_0$

Since the incident proton's total energy E_i is the sum of its kinetic energy $(E_k)_i$ and rest energy E_0,

[28-19] $(E_k)_i = E_i - E_0 = 7E_0 - E_0 = 6E_0$

The incident proton's kinetic energy at the threshold is *six* times the proton rest energy, whereas one might imagine only *two* rest energies would be required to bring a proton and antiproton into existence. Actually a large fraction of the incident particle's kinetic energy is wasted in putting the emerging four particles into motion, in order that momentum be conserved. Contrast this situation with a symmetrical collision of two protons, where the total momentum of the system before collision is zero (the center of mass is at rest in the laboratory) and the four particles appearing after the collision may be at rest. In fact, a total kinetic energy of $2E_0$ before collision is just enough to create the additional proton and antiproton. In short, whereas protons with kinetic energy of at least $6E_0$ are required to produce antiprotons when striking protons at rest, protons with kinetic energy of only $1E_0$ can produce antiprotons when colliding with similar, oppositely directed protons.

For a given energy input, then, much more energetic particles can be created when collisions are symmetrical in the laboratory. To exploit this fact, some accelerating machines use two beams of particles which collide head on. Alternatively, the single beam of an accelerator can be stored in a ring and later directed at particles moving in the opposite direction (see Fig. 28-7).

Colliding beams

28-5 BETA DECAY REVISITED

Unlike the proton, the neutron is unstable. In Section 27-5 it was found that an unstable nucleus may decay with the emission of a beta electron, the simultaneous emission of an antineutrino, and the concomitant conversion of one nuclear neutron into a proton. For example,

$$^{12}_{5}B \rightarrow {}^{12}_{6}C + {}^{0}_{-1}e + {}^{0}_{0}\bar{\nu}$$

The fundamental beta-decay process, of which this is but one example, is

$$n \rightarrow p + e + \bar{\nu}$$

By comparing rest masses it is easy to verify that the last process is energetically allowed and that the total kinetic energy of the emergent particles is 0.78 MeV. Decay of a *free* neutron, not merely one within a nucleus, has been observed directly; it takes place relatively slowly, the mean life of the neutron being 9.3×10^2 s.

That an *antineutrino* is given off in beta decay can be deduced from the conservation laws of baryon number and electron number:

An *antineutrino* is emitted in the decay of a neutron.

$$n \longrightarrow p + e + \bar{\nu}$$ [28-20a]

Baryon number: $(+1) = (+1) + 0 + 0$
Electron number: $0 = 0 + (+1) + ?$

We see that these conservation laws are satisfied only if the massless particle (hence clearly not a baryon) is assigned an electron number of -1 corresponding to an antiparticle, the antineutrino ($+1$ corresponding to a neutrino).

Recall now that in any decay we can reverse the arrow (imagine

FIGURE 28-7 (*a*) Simple arrangement for two beams colliding head on. (*b*) The $80-million CERN (Center for European Nuclear Research) installation at the Swiss-French border (dashed line) utilizes the ISR (Intersecting Storage Ring) principle of (*a*) to produce colliding 28-GeV beams. (*Photo CERN.*)

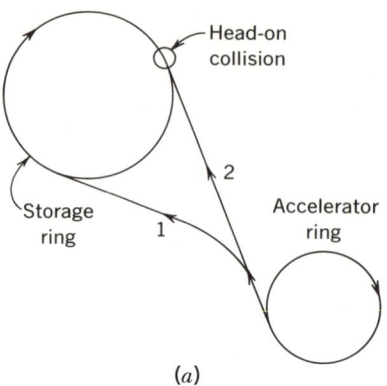

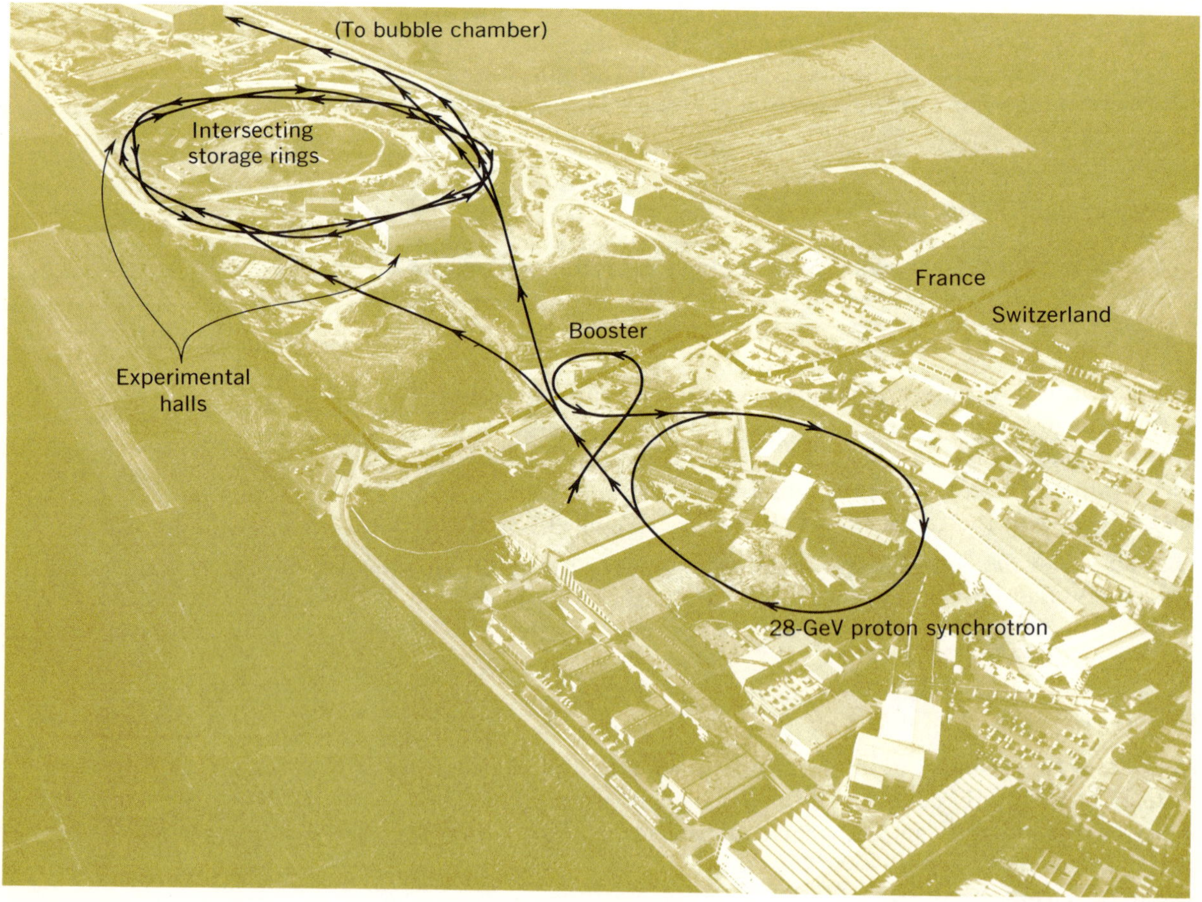

the process as going backward in time). Furthermore, we can shift a particle from one to the other side of the reaction equation if at the same time we change it into its antiparticle. By means of these two operations we can transform

$$n \to p + e + \bar{\nu}$$

to

$$n \leftarrow p + e + \bar{\nu}$$

to

$$n + \bar{e} + \nu \leftarrow p$$

or

$$p \to n + \bar{e} + \nu \qquad [28\text{-}20b]$$

obtaining as a final result the possible decay of a proton into a neutron, positron, and neutrino. For a free proton, outside a nucleus, this process is forbidden by the energy-conservation law: the proton rest mass is *less* than that of neutron and positron (see Example 28-3); but a proton bound to other neutrons and protons in an unstable nucleus can undergo such decay. The required energy is borrowed, so to speak, from neighboring nuclear particles with which the decaying proton interacts. For example, we may have beta decay with the emission of a positron through the decay of nitrogen 12 nuclei

Positron emission

$$^{12}_{7}\text{N} \to {}^{12}_{6}\text{C} + {}^{0}_{+1}e + {}^{0}_{0}\nu$$

EXAMPLE 28-3

(*a*) What is the total kinetic energy released when a neutron decays into a proton, electron, and antineutrino? (*b*) A neutron can decay into a proton, but a proton cannot decay into a neutron. Why?

SOLUTION

a Taking the decaying neutron to be initially at rest in the process

$$n \to p + e + \bar{\nu}$$

the total energy of the system initially is simply the rest energy of the neutron, $E = m_0 c^2 = (1.008665 \text{ u})c^2$. The rest energies of the other particles can similarly be given as their rest masses multiplied by the constant c^2. Therefore, the total *rest* energies of the particles emerging from the decay are

Proton	$(1.007277 \text{ u})c^2$
Electron	$(0.000549 \text{ u})c^2$
Antineutrino	0
Total *rest* energy of decay particles	$(1.007826 \text{ u})c^2$

The total energy into the decay is the rest energy of the neutron. The total energy out of the decay is the rest energy of the particles emitted together with their total kinetic energy. Therefore, the kinetic energy of the proton, electron, and antineutrino is

$$E_k = (1.008665 \text{ u})c^2 - (1.007826 \text{ u})c^2$$
$$= (0.000839 \text{ u})c^2 \times [931.5 \text{ MeV}/(\text{u})(c^2)] = 0.78 \text{ MeV}$$

b Decay of a neutron into a proton, electron, and antineutrino is energetically allowed because the rest mass of a neutron exceeds that of the proton, electron, and antineutrino. Clearly, then, an isolated proton cannot decay into a neutron because its rest mass is less than that of a neutron.

Altering the basic neutron decay relation again, so that the proton and electron remain on one side, we have

$$n \rightarrow p + e + \bar{\nu}$$
$$n + \nu \leftarrow p + e$$

or

[28-20c] $\quad p + e \rightarrow n + \nu$

Here a proton captures an electron to produce a neutron and neutrino. This process cannot occur for a free proton, again because it is disallowed by energy conservation; but, again, a proton in an unstable nucleus can seize one of the nearby atomic electrons, a process known as *electron capture*. One example is

Electron capture

$$_{-1}^{0}e + {}_{4}^{7}\text{Be} \rightarrow {}_{3}^{7}\text{Li} + {}_{0}^{0}\nu$$

in which radioactive beryllium decays into lithium. The sole energetic particle emitted in the reaction is the elusive neutrino. That the lithium 7 nucleus is actually observed to recoil is strong evidence for the existence of the neutrino. The most emphatic demonstration, however, comes from a reaction which the neutrino initiates. Starting once again with the basic neutron-decay relation, and now arranging the particles so that an antineutrino is absorbed by a proton, we have

Antineutrino capture

$$n \rightarrow p + e + \bar{\nu}$$
$$n + \bar{e} \leftarrow p + \bar{\nu}$$

or

[28-20d] $\quad p + \bar{\nu} \rightarrow n + \bar{e}$

This process has been recorded in the core of a nuclear reactor. So weak is the interaction of neutrinos or antineutrinos with other matter that an enormous number of them is required to result in just one observable event. A nuclear chain reaction is a prolific source of neutrinos: each fission event results in unstable nuclei, most of which decay, emitting electrons and antineutrinos. To show that a proton has absorbed an antineutrino one must show that a positron and a neutron are produced simultaneously. The positrons are identified by looking for the 0.51-MeV photons that appear when a positron annihilates with an electron. The neutrons are soon absorbed in the surrounding material, producing unstable nuclei which shortly thereafter may undergo gamma decay. As a result, confirmation of the antineutrino absorption by protons is made by finding 0.51-MeV photons closely followed by gamma decays.

28-6 CONSERVATION OF MUON NUMBER

Pion and muon decay

Beta decay of nuclei is not the only source of neutrinos and antineutrinos, which also appear in the decay of other unstable particles. For example, charged pions, which may be produced when highly energetic protons collide with matter, are very unstable. A charged pion quickly (with a mean life of 2.6×10^{-8} s) decays into an antimuon or muon and a neutrino or antineutrino

$$\pi^+ \to \bar{\mu} + \nu \quad [28\text{-}21a]$$

$$\pi^- \to \mu + \bar{\nu} \quad [28\text{-}21b]$$

where $\bar{\mu}$ is *positively* charged and μ is *negatively* charged.

Muons, too, are unstable. The positive muon decays (mean life 2.2×10^{-6} s) into a positron and two massless particles, a neutrino and antineutrino, and the negatively charged muon decays, as shown in Fig. 28-8, into an electron, together with a neutrino and antineutrino

$$\bar{\mu} = \bar{e} + \nu + \bar{\nu} \quad [28\text{-}22a]$$

and

$$\mu = e + \nu + \bar{\nu} \quad [28\text{-}22b]$$

That a muon decays into *three* particles, rather than just two, follows from the same kind of observations that originally revealed the neutrino in beta decay. It is found that the electrons and positrons produced in the decay of muons do not recoil with a single kinetic energy but have a distribution of energies that indicates two accompanying particles, each with a zero rest mass.

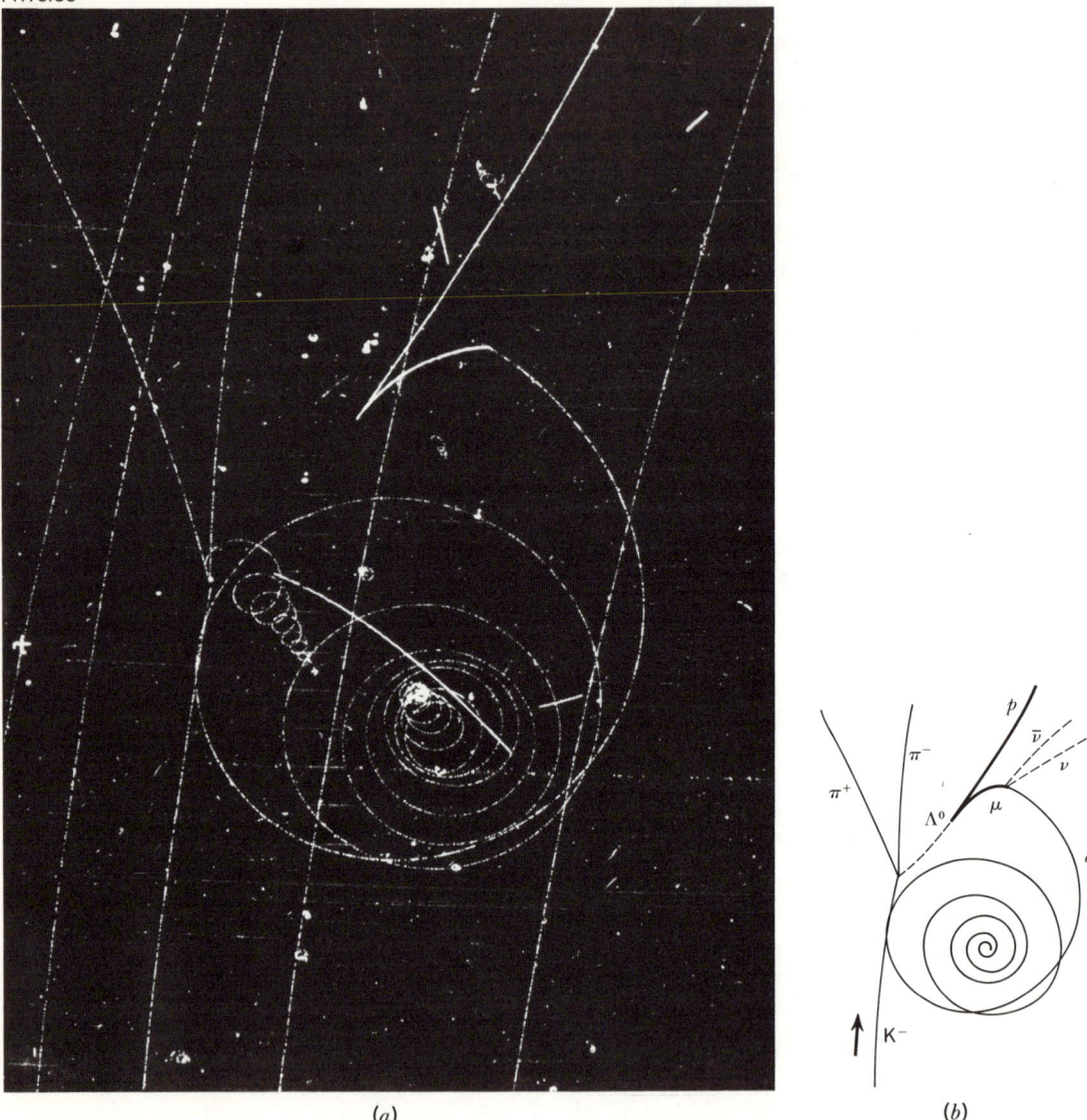

FIGURE 28-8 Photograph showing the decay of a muon into an electron: $\mu \to e + \nu + \bar{\nu}$. (*University of California, Lawrence Radiation Laboratory, Berkeley.*)

One might suppose the neutrinos and antineutrinos emitted in the decay of pions and of muons to be identical with the neutrinos and antineutrinos appearing in beta-decay processes. Thus, we would suppose all neutrinos and antineutrinos to belong to the electron family of particles, carrying electron number $+1$ (neutrino) or -1

(antineutrino). Furthermore, the antineutrino from pion decay also ought to initiate the reaction [28-20d]

$$\bar{\nu} + p \to n + \bar{e}$$

for we have seen that this reaction furnishes positive identification of the antineutrino produced in beta decay.

But the experiments tell *otherwise*. When an antineutrino from a π^- decay is captured by a proton, the result is a neutron and a positively charged *muon*

$$\bar{\nu} + p \to n + \bar{\mu} \qquad [28\text{-}23]$$

In summary, for the antineutrinos emitted in the beta-decay process,

$$n \to p + e + \bar{\nu}$$
$$\bar{\nu} + p \to n + \bar{e} \quad (\text{not } n + \bar{\mu})$$

whereas for the antineutrinos emitted with muons in the decay of pions

$$\pi^- \to \mu + \bar{\nu}$$
$$\bar{\nu} + p \to n + \bar{\mu} \quad (\text{not } n + \bar{e})$$

We can only conclude that the antineutrinos (and neutrinos) given off by decaying pions are fundamentally different from those emitted in beta decay. Although they have zero charge and zero rest mass, the antineutrinos and neutrinos produced by decaying pions cannot be members of the electron family. In fact, they belong to the *muon* family, and are assigned muon numbers.

<small>Neutrinos (and antineutrinos) of two different kinds: those from beta decay, members of the electron family; those from pion decay, members of the muon family</small>

One must pause in admiration. Here are particles without charge, without rest mass, things so fugitive that they have an even chance of penetrating several light-years of solid lead. Yet now these shadowy entities are distinguished into two kinds!

As always, the importance of assigning a particle to a family lies in the fact that the corresponding number is conserved in all processes. More specifically, the total muon number is unchanged in any process if the (negative) muon and the (positive) antimuon are arbitrarily given muon numbers of $+1$ and -1, respectively, and the associated neutrino and antineutrino are given muon numbers, respectively, of $+1$ and -1. Hereafter, neutrinos and antineutrinos of the muon family are denoted ν_μ and $\bar{\nu}_\mu$ and those of the electron family ν_e and $\bar{\nu}_e$.

Recognizing, then, the existence of three families of elementary particles, the electron family, the baryon family, and now the muon

CHAPTER 28 ELEMENTARY-PARTICLE PHYSICS

Conservation law:	Observed Process					
	$\pi^+ \to \bar{\mu} + \nu_\mu$	$\pi^- \to \mu + \bar{\nu}_\mu$	$\bar{\mu} \to \bar{e} + \nu_e + \bar{\nu}_\mu$	$\mu \to e + \bar{\nu}_e + \nu_\mu$	$\bar{\nu}_e + p \to n + \bar{e}$	$\bar{\nu}_\mu + p \to n + \bar{\mu}$
Electron number	$0 = 0 + 0$	$0 = 0 + 0$	$0 = (-1) + (+1) + 0$	$0 = (+1) + (-1) + 0$	$(-1) + 0 = 0 + (-1)$	$0 + 0 = 0 + 0$
Muon number	$0 = (-1) + (+1)$	$0 = (+1) + (-1)$	$(-1) = 0 + 0 + (-1)$	$(+1) = 0 + 0 + (+1)$	$0 + 0 = 0 + 0$	$(-1) + 0 = 0 + (-1)$
Baryon number	$0 = 0 + 0$	$0 = 0 + 0$	$0 = 0 + 0 + 0$	$0 = 0 + 0 + 0$	$0 + (+1) = (+1) + 0$	$0 + (+1) = (+1) + 0$

TABLE 28-3

family, we rewrite in Table 28-3 the previously given reactions with all family numbers assigned.

28-7 THE FUNDAMENTAL INTERACTIONS

In discussing the conservation of electron family number we noted that the uncertainty principle opened up a radical possibility. Violations of the conservation laws may occur, but unobservable ones! The present section will show that such "subliminal" events figure essentially in the quantum theory of interparticle forces.

We begin by reconsidering the two basic electron-photon interactions. In emitting a photon, an electron has its momentum and energy changed, and we write

$$e_1 \to e_1' + \gamma \quad \text{near another particle}$$

where e_1 and e_1' symbolize, respectively, the electron before and after the photon emission. We recognize that a single free electron in empty space cannot actually undergo this process because it would violate the conservation laws of energy and momentum.

The absorption of a photon by an electron can be written

$$\gamma + e_2 \to e_2' \quad \text{near another particle}$$

where the momentum and energy of the electron are now changed by the absorption of the photon. Once again, a single free electron in empty space cannot undergo the process because it too would violate the laws of energy and momentum conservation.

Suppose, however, the two processes are taken together: one electron spontaneously emits a photon which is later absorbed by a second electron. Symbolically,

$$[28\text{-}24] \quad e_1 + e_2 \to (e_1' + \gamma) + e_2 \to e_1' + (\gamma + e_2) \to e_1' + e_2'$$

The process is shown diagrammatically in Fig. 28-9. A photon has been exchanged between the two electrons, and as a consequence the momentum and energy of each of the electrons has been changed. In effect, each electron has been acted upon by *a force*. We are thus led to a new interpretation of the interaction between a pair of electrically charged particles: it consists in the creation, trading, and annihilation of photons. This quantum theory of the electromagnetic field is similar to the classical theory in that it assigns the creation of photons to electric charges,[1] but the new theory is more mechanical in regard to the electromagnetic field itself. No longer an impalpable something which arises from one charge and operates at the site of another, the field is now a swarm of energy- and momentum-carrying photons which pop in and out of existence.

Although the energy- and momentum-conservation laws are satisfied neither in the spontaneous creation of a photon by one charged particle nor in the spontaneous absorption of the photon by another charged particle, these fundamental principles are *not* violated in the overall process, extending in Fig. 28-9 from point A to point B, which corresponds to the time during which the photon is in existence. To see this we recall an implication of the Heisenberg uncertainty relations [25-11] and [25-12]

$$\Delta p_x \, \Delta x \gtrsim h$$
$$\Delta E \, \Delta t \gtrsim h$$

If a particle's momentum (along x) is uncertain by an amount Δp_x, the particle may be localized (along x) only with an uncertainty $\Delta x \gtrsim h/\Delta p_x$. Similarly, if a particle's energy is measured over some finite time interval Δt, that energy must be uncertain by at least an amount $\Delta E \approx h/\Delta t$. If a particle's momentum is uncertain to some degree, this implies that the law of momentum conservation may actually be violated by an amount Δp_x provided that the violation takes place over a region of space at most as large as $\Delta x \approx h/\Delta p_x$. And, if a particle's energy is uncertain, the law of energy conservation may be violated by an amount ΔE, provided that the period Δt of nonconservation of energy is no greater than $h/\Delta E$. In short, the uncertainty relations allow momentum and energy to be borrowed so long as the debt is repaid within the limits imposed by the size of the constant h.

Returning now to the photon exchange shown in Fig. 28-9, we see that if the momentum and energy borrowed at point A are repaid at point B, all within the limitations of the uncertainty principle, then energy and momentum are indeed conserved for the overall

A photon exchanged between electrically charged particles: an electromagnetic force

[1] Recall that, according to Maxwell, it is always the acceleration of electric charges that generates light waves.

Violations of momentum and energy conservation allowed by the uncertainty principle

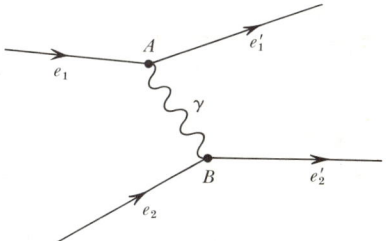

FIGURE 28-9 Mutual interaction of two electrons: electron e_1 spontaneously emits a photon γ at point A which is absorbed by electron e_2 at point B.

process. When points A and B are close together, the photon, traveling the short distance at the speed of light, exists only for a short time. With both Δx and Δt very small, the momentum and energy discrepancies Δp_x and ΔE may be substantial, so that the photon's momentum and energy are relatively large. A large transfer of momentum over a small time interval means a large force acting between the two charged particles. On the other hand, if A and B are far separated, the photon's energy and momentum must be correspondingly smaller. Since the photon has zero rest energy, the minimum value of the energy which must be borrowed may be as small as zero. Thus ΔE must extend to zero; and Δt, and hence also Δx, extend to infinity. There is no spatial limit to the electromagnetic interaction. Two electrons at a small separation will interact strongly via high-energy photons, and two electrons at a large separation will interact weakly via low-energy photons. This corresponds exactly to the fact that the ordinary Coulomb force (as well as the magnetic force), which decreases inversely with the square of the distance between the charged particles, drops to zero only for an infinite separation distance.

Infinite-range electromagnetic force corresponds to zero rest mass for photon.

The photons responsible for the electromagnetic force between charged particles are said to be *virtual photons,* so named because they cannot be observed directly. The uncertainty principle guarantees that whatever their energy, they will not last quite long enough to leave a trace.

Although Fig. 28-9 shows two electrons interacting through the exchange of a virtual photon, any two electrically charged particles can interact in the same fashion. The electromagnetic interaction between a pair of protons is diagrammed in Fig. 28-10.

The two protons repel each other because both carry an electric charge and so are capable of creating and annihilating the particle of the electromagnetic interaction, the photon. But we know that a pair of protons separated by a distance of no more than about 10^{-15} m (approximately the separation between particles within an atomic nucleus) attract each other strongly through the nuclear force. The strong nuclear interaction is also manifested between a pair of neutrons or between a proton and neutron inside a nucleus.

FIGURE 28-10 The electromagnetic interaction between a pair of protons: proton p_1 spontaneously emits a photon γ, which is absorbed by proton p_2.

If the electromagnetic interaction can be described in terms of the exchange of a particle (the photon) between electrically charged particles, we should expect that the nuclear interaction can also be described in terms of the exchange of a particle. A process by which a pair of protons interact through the nuclear interaction can be

SECTION 28-7
THE FUNDAMENTAL INTERACTIONS

Nuclear force through the exchange of pions

written

$$p_1 + p_2 \rightarrow (p_1' + \pi°) + p_2 \rightarrow p_1' + (\pi° + p_2) \rightarrow p_1' + p_2' \qquad [28\text{-}25]$$

(see Fig. 28-11). The exchange particle is the electrically neutral pion (pi meson), which is created spontaneously at point A by the first proton and annihilated spontaneously at point B by the second proton. As before, both the momentum and energy conservation laws are violated separately at points A and B, but there is no net violation in the overall process. The exchange particle for the nuclear force differs from that for the electromagnetic force in one important respect: whereas a photon may have any energy (because its rest energy is zero), the pion has a minimum energy corresponding to its rest energy of about 140 MeV.

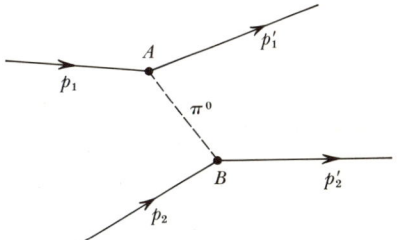

FIGURE 28-11 The nuclear interaction between a pair of protons; the two particles trade a pion.

It is this difference between photon and pion which accounts for the difference in range between the electromagnetic and strong nuclear interactions. When a neutron or proton creates a virtual pion, the minimum amount by which the energy-conservation principle is violated is the rest energy of about 140 MeV. This imposes, through the uncertainty relation, a limitation on the time during which the pion may exist and the distance it may travel before being absorbed by a second proton or neutron. Indeed, the uncertainty principle can be used to show (see Problem 28-6) that the exchange particle associated with the strong nuclear interaction must have a finite rest energy of the order of 140 MeV if the range of the nuclear force is to be approximately 10^{-15} m![1]

Although the photons and mesons whose exchange accounts for the electromagnetic and strong nuclear interactions are unobservable, or virtual, we can, of course, observe real photons and real mesons if no energy or momentum debt need be repaid. For example, doing work on and accelerating a charged particle will cause it to emit electromagnetic radiation; in the language of quantum physics, supplying energy to a charged particle will allow it to create real photons. Similarly, mesons are brought into real existence by nuclear interactions, as in the mutual annihilation of a proton and antiproton

$$p + \bar{p} \rightarrow \pi^+ + \pi^- + \pi° + \pi°$$

This last reaction exhibits all three varieties of pion: $\pi°$, π^+, and π^-, with electric charges, respectively, of 0, $+e$, and $-e$. All three have very nearly the same rest mass. Like the neutral pion, the charged pions, as virtual particles, can serve as the field quanta for the nuclear interaction. For instance, a proton and a neutron can exchange a

---

[1] By arguments of the sort just given H. Yukawa was able to predict that because of the short-range character of the nuclear force, particles of mass intermediate between electrons and protons (mesons) should exist. This he did in 1935, or 12 years before actual observation of mesons in the laboratory.

virtual π^+ and thereby trade identities (Fig. 28-12)

[28-26] $\quad p + n \rightarrow (n + \pi^+) + n \rightarrow n + (\pi^+ + n) \rightarrow n + p$

Since no family conservation law applies to pions (the same is true of photons), their total number in any process is limited only by the remaining conservation laws.

Because the electromagnetic interaction is mediated by photons and the nuclear interaction by mesons, it has been conjectured that the gravitational interaction, too, has its characteristic particle (Fig. 28-13). This particle, which has not yet been observed, has been given the name *graviton*. We know that the graviton, like the photon must have zero rest energy and travel at the single speed c, for the gravitational interaction, like the electromagnetic interaction, has an infinite range. The difficulties in detecting gravitons are connected with the extreme feebleness of the gravitational force, as compared with the electromagnetic or nuclear force. One must first establish the existence of gravitational radiation; the results of some recent experiments suggest this. One might expect to find that a gravitational disturbance (say, the sudden explosion of a star) propagates at the finite speed c. Then one must show that gravitational radiation is quantized into gravitons, just as electromagnetic radiation is quantized into photons. In view of the smallness of the gravitational constant G, this remains at present a remote possibility.

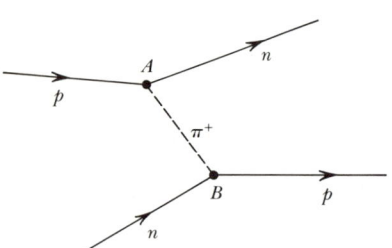

FIGURE 28-12 The nuclear interaction between a proton and neutron. When an electrically charged pion is traded between the two particles, their identity is interchanged.

In addition to the electromagnetic, strong (or nuclear), and gravitational interactions, there is a fourth fundamental interaction, known merely as the *weak interaction* because the associated forces are weak compared to those of the strong interaction. It is, however, stronger than the gravitational interaction. The weak interaction is illustrated in the fundamental beta-decay process

$$n \rightarrow p + e + \bar{\nu}_e$$

where a neutron decays spontaneously into a proton, electron, and antineutrino (of the electron family type). Equations [28-21a] and [28-21b] are examples of spontaneous decay through the weak interaction, as is the decay of positive or negative pions into muons and neutrinos (of the muon family).

Figure 28-14 illustrates the weak interaction of a neutron and a proton. The neutron undergoes beta decay, emitting an electron and an antineutrino and becoming a proton. The primary proton absorbs the electron and the antineutrino, thereby becoming a neutron. (The conjectured intermediate particle for the weak interaction is sometimes called the W particle.)

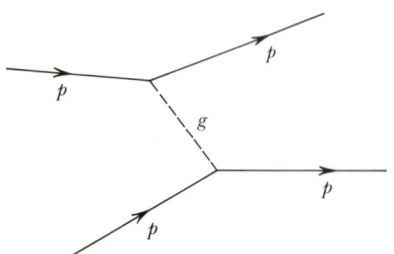

FIGURE 28-13 The gravitational interaction between a pair of protons; it has been conjectured that the two particles trade a graviton g.

Fundamental Interaction	Relative Strength	Field Particle
Strong	1	Pion
Electromagnetic	10^{-2}	Photon
Weak	10^{-13}	W particle (?)
Gravitational	10^{-40}	Graviton (?)

TABLE 28-4

Table 28-4 summarizes the four fundamental interactions. All processes involving the elementary particles—decay, absorption, and the interchange of particles—can be identified with one or more of these fundamental interactions.

The lifetime of an elementary particle is a measure of the strength of the interaction through which it decays. Particles that decay by strong interactions have mean lives of about 10^{-20} s, while electromagnetic decays are about 10^{-15} s. Weak-interaction decays are typically 10^{-10} s. The proton and electron have infinite lifetimes (strictly, greater than 2×10^{28} years and 2×10^{21} years by indirect measurement) and are completely stable; their decay is precluded by the conservation laws of baryon and electron family number.

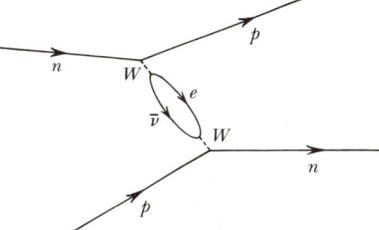

FIGURE 28-14 The weak interaction between a proton and neutron; it is conjectured that the proton and neutron trade a W particle which is thought of as an intermediate particle in [28-20].

The seven conservation laws listed in Section 28-1[1] have been found, at least up to this time, to be universally valid; they describe *all* the observed interactions of elementary particles, whether through the strong, the electromagnetic, the weak, or the gravitational force. In recent years a number of additional conservation laws have been discovered which are not universally valid. They describe *some* interactions but not all. Here we shall consider one example, the law of parity conservation, which is found to apply to all interactions except the weak. We shall divide this topic into three parts: (1) a discussion of the spin angular momentum of elementary particles, (2) a description of parity conservation and other related principles, and (3) an illustration of parity nonconservation in the weak interaction.

1 *Spin angular momentum* Imagine that a space platform consists of two identical disks both mounted freely on the same axle, as shown in Fig. 28-15a. Two astronauts of equal mass are attached, one to each disk. We know from (linear) momentum conservation what happens if the astronauts push each other in directions parallel to the axle: the identical disks recoil in opposite directions with equal speeds. The (linear) momentum of the system has a value that is always equal to its initial value, zero.

28-8 THE NONCONSERVATION OF PARITY IN WEAK INTERACTIONS

[1] Momentum, mass-energy, electric charge, angular momentum, electron number, muon number, and baryon number.

Spin angular momentum and a right-handed object

CHAPTER 28 ELEMENTARY-PARTICLE PHYSICS

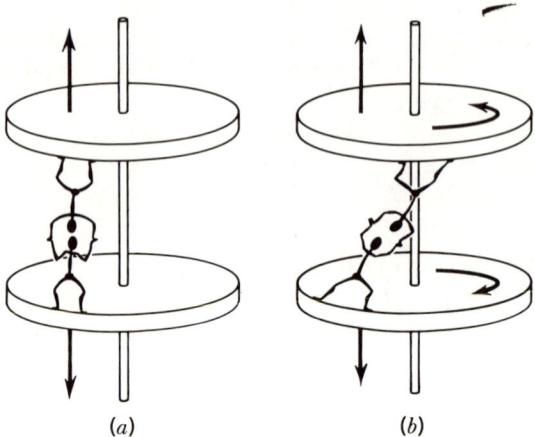

FIGURE 28-15 Two identical disks in empty space with two astronauts: (*a*) when the two astronauts push in opposite directions parallel to the axis, the disks recoil in opposite directions with equal speeds; (*b*) when the astronauts push each other with forces having components tangential to the edge of the disks, in addition to recoiling in opposite directions, the disks rotate in opposite senses with equal angular speeds.

If the astronauts are attached near the edges of the disks and push each other with forces having components tangential to the edges of the disks (Fig. 28-15*b*), the disks will also spin about the axle as they recoil from each other. We know that identical disks will rotate with equal angular speeds, one clockwise and the other counterclockwise.

Angular momentum, as shown in Chapter 7, is a measure of the distribution and rotation of inertial mass about some axis. It is an especially useful physical quantity because the total angular momentum of an isolated system, like its total (linear) momentum, is conserved. In the above example, there is no initial rotation, and the initial angular momentum of the system is zero. The final value of the total angular momentum is also zero because the negative angular momentum associated with the clockwise rotation of one disk is perfectly balanced by the positive angular momentum associated with the counterclockwise rotation of the other identical disk. The angular momentum of the system has a value that is always equal to its initial value, zero.

Objects, like the disks that are symmetrical about their axis of rotation have angular momentum that is independent, both in magnitude and sign, of the choice of inertial frame and the choice of axis for calculating the angular momentum (Section 7-2). This invariant portion of their angular momentum is called *spin angular momentum*.

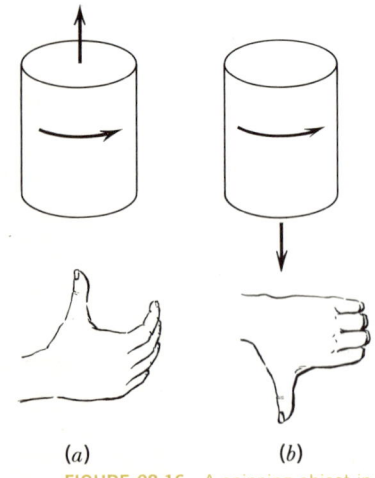

FIGURE 28-16 A spinning object in motion along the line of its spin axis: (*a*) a right-handed and (*b*) a left-handed spinning object.

Consider a spinning object which is in motion along a line corresponding to its spin axis, as shown in Fig. 28-16*a*. The object

+ SECTION 28-8
THE NONCONSERVATION OF PARITY IN WEAK INTERACTIONS

can be characterized as being right-handed, for when the curled fingers of the right hand correspond to the spin sense, the right thumb gives the direction in which the object is traveling. Figure 28-16*b* shows a spinning object which can be said to be left-handed; it spins in the same sense as the one shown in Fig. 28-16*a*, but it travels in the opposite direction and its spin sense and direction of motion are related in the same fashion as the curled fingers and thumb of the *left* hand.

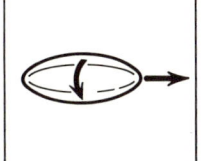

FIGURE 28-17 A right-handed object: a spinning football. View the figure through the reverse side of the page to see how the football becomes left-handed.

It is a simple matter to see that a right-handed spinning object can become a left-handed spinning object simply by making a different choice of the reference frame in which we view the motion. Suppose that we view the spinning object shown in Fig. 28-16*a* from some initial reference frame; as the object moves upward, it is right-handed. Now suppose that we view the spinning object from another reference frame, one traveling upward relative to the initial reference frame at a speed exceeding that of the spinning object. From the point of view of an observer in this new reference frame, the object travels downward. On the other hand the spin sense has remained unchanged. In effect, the initially right-handed object has become a left-handed object through a change in reference frame.

A useful way to think of the relationship between a right- and a left-handed object is in terms of a motion-picture film. If a properly run film shows a right-handed object, the same film inserted in the projector with sides reversed will show a left-handed object. Figure 28-17 shows a football thrown as a right-handed object; to observe how it would look when inserted in a projector with sides reversed, look at this same figure through the *back* of the page.

Query If you have thrown a football, explain why a spinning football thrown overhand by a right-handed player is a right-handed object while one thrown overhand by a left-handed player is a left-handed object. Can a right-handed player throw the football so that it will be a left-handed object?

Another way of holding this relationship in mind is to recollect that the mirror image of a right-handed object is left-handed and vice versa (look back at Fig. 17-3). Figure 28-18 shows the mirror image of the football shown in Figure 28-17. The football seen in the mirror is the same as the one seen before through the back of this page.

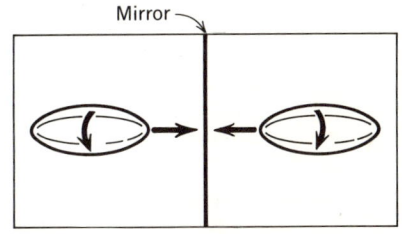

FIGURE 28-18 The mirror image of a right-handed spinning object is a left-handed spinning object.

All the foregoing remarks about spin angular momentum and handedness apply to ordinary objects of finite size. They also apply to single elementary particles, i.e., particles carrying certain properties which cannot be divided or separated from the particle. In general, the *total* angular momentum of a particle depends on the particular choice of the inertial frame and the axis used for the computation. But an electron (as well as other elementary particles) is found always to have some angular momentum which is independent of the choice of inertial frame and axis. This is its spin angular momentum. The spin angular momentum of an elementary particle, like its charge, is an intrinsic property in that it is unaffected by any external influences that act on the particle. Indeed, it is by virtue of having certain properties which cannot be divided or separated away that a particle is considered as elementary. Since we never consider a particle as having internal structure, it makes no sense to speak about the internal rotation of a particle about some axis; there is no way (and no need) to explain the spin angular momentum of elementary *particles*. Nevertheless, we sometimes think of electrons and other elementary particles not as particles but as tiny perpetually spinning balls, just for the sake of having some kind of mental picture.

Spin angular momentum, an intrinsic property of an elementary particle

It can be shown that the *orbital* angular momentum (component) of hydrogen is always an integral multiple of $\hbar$, Planck's constant divided by 2π. Not only is orbital angular momentum conserved; according to the quantum theory, it is, like energy, quantized. However, it turns out that a component of the *spin* angular momentum of each variety of elementary particle is always an integral multiple of $\frac{1}{2}\hbar$. For convenience we take $\hbar$ as the unit of measurement for spin angular momentum, and for short, we say the electron has spin $\frac{1}{2}$. Likewise, both the muon and neutrino have spin $\frac{1}{2}$. But the pion has spin 0 and the photon spin 1. Values for still other elementary particles are listed in Table 28-1.

Spin-$\frac{1}{2}$ particles are found always to have their spin axis along their direction of motion. This means that they can be characterized as being either right- or left-handed, the assignment for a given particle of finite rest mass depending on the choice of reference frame.

Time T reversal: run the movie backward in time

2   *TCP reversal*   Imagining a movie run backward in time helped us appreciate that an allowable process on the particle level remains allowable, i.e., still in accord with the same laws of physics, when it proceeds the other way in time.[1] Our last application of this idea was in connection with the various guises of the electron-photon interaction. In formal terms we shall say that the laws of physics are invariant under time reversal.

Keeping to the movie metaphor, let us next imagine that a film of colliding billiard balls is projected with sides reversed (back-

[1] Of course, we would be able to tell the difference between a film run forward or backward in time if it portrayed some series of events involving macroscopic objects that consist of *many* interacting particles, e.g., an egg striking a pavement or a diver entering a pool. But at the microscopic level, where one deals ultimately with interactions between *pairs* of particles, the arrow of time can be reversed, as we discussed in Chapter 11.

ward in left-right space). The new billiard-ball collision, which differs from the old one only in handedness, is again in accord with the laws of physics. The operation of interchanging left and right is referred to as *parity inversion*. Because nature apparently shows no preference between left and right, either a direct view or a reversed view of events in nature is allowable. The same physical laws are followed. In formal terms we shall say that the fundamental laws of physics are invariant under parity inversion; equivalently, we say parity is conserved.

Parity P reversal: reverse movie film left-right

Imagine, finally that we have a motion picture of two interacting electrically charged particles, say an electron and a positron. We might add the identifying labels minus and plus to the images of the two particles on the film. As before, if we run the film backward in time or reverse it left-right, the series of events which unfolds is describable in terms of the very same laws of electromagnetic interaction. Moreover, if we interchange the labels plus and minus, the events which now unfold "backward in charge" are still in accord with the laws of electromagnetism. In reversing the plus and minus originally assigned to the positron and electron, respectively, we have in effect converted an antiparticle to a particle and a particle to an antiparticle. The interchange of particle and antiparticle produces no new phenomena. In formal terms we say that the laws of physics are invariant under charge reversal.

Charge C reversal: interchange plus and minus (particle and antiparticle)

The laws of reversibility of time T, parity P, and charge C can be applied to elementary events singly or in combination. Thus, one can reverse the direction of the film's motion through the projector (T), or reverse the film left-right (P), or interchange the labels for particle and antiparticle (C), or perform any two or all three of these operations and expect to find the new events fully in accord with physical laws. It seems self-evident that nature, on a microscopic scale, does not prefer before over after, or left over right, or positive over negative charge (particle over antiparticle).

It seems self-evident—but it is false. One of the most remarkable developments in recent physics[1] was to show that in events governed by the weak interaction, parity is not conserved! Nature, one might say, is not evenhanded. Whether the remaining invariance principles of time-reversibility and of particle-antiparticle reversibility are also violated in some interactions is under intensive investigation at present.

[1] This won Yang and Lee the Nobel Prize in physics in 1957. Lee was twenty-nine and Yang thirty-three.

3 *Parity nonconservation in the decay of a pion* The simplest example of parity nonconservation arises in the decay of a pion into a muon and mu neutrino

$$\pi^+ \rightarrow \bar{\mu} + \nu_\mu$$

or

$$\pi^- \rightarrow \mu + \bar{\nu}_\mu$$

(Note that, among others, the conservation laws of electric charge and of muon family number are satisfied in these decays.) The mean life of the π for decay by this process is 2.6×10^{-8} s. This time is relatively long compared to that associated with decay by the electromagnetic interaction ($\sim 10^{-15}$ s) or by the strong interaction ($\sim 10^{-20}$ s), revealing that the pion decay into a muon and neutrino is governed by the weak interaction.

Figure 28-19a shows a π^+ initially at rest decaying into a $\bar{\mu}$ and a ν_μ. Since the pion has zero momentum, the two decay particles must fly off in opposite directions. Moreover, since the spin angular momentum of the pion is zero (Table 28-1), the spin angular momenta of the two departing particles must be equal and opposite: if one spins clockwise, the other must spin counterclockwise. There appear to be two distinct possibilities, then: (1) the muon right-handed and the neutrino also right-handed, as shown in Fig. 28-19b or (2) the muon left-handed and likewise the neutrino, as seen in Fig. 28-19c. Since the conditions before the decay have perfect right-left symmetry (remember the pion itself has no spin), either right-handed products or left-handed products are to be expected with equal probability. Therefore, starting with a large collection of identical pions at rest initially, one expects on the basis of left-right invariance (parity conservation) to find equal numbers of left- and right-handed muons appearing in the decays.

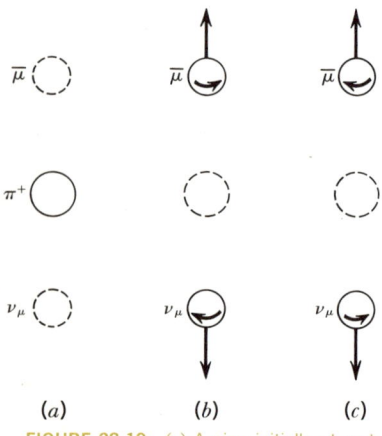

FIGURE 28-19 (*a*) A pion initially at rest decays into a muon and neutrino with two possibilities: (*b*) muon and neutrino both right-handed or (*c*) muon and neutrino both left-handed.

The experimental fact, however, is that only the decay shown in Fig. 28-19c ever takes place; left-handed, but *never* right-handed, muons are observed.[1] In this weak interaction left-right invariance is violated; i.e., there is a preference for left-handedness, and parity is not conserved. Evidently, what is true for fundamental laws describing the large-scale world is not true for the laws that describe the interaction of these elementary particles. Parity conservation is not a *universal* law.

How can we account for the exclusive appearance of left-handed muons and neutrinos when a π^+ decays? Could it be that right-

---

[1] In principle one can establish the handedness of muons in motion by observing the direction of recoil and the sense of rotation of an object absorbing a muon; in actuality this procedure is extraordinarily difficult, and the handedness of muons is determined by more subtle means. Moreover, because neutrinos pass through materials virtually without absorption, one cannot measure the handedness of a neutrino directly. However, using angular-momentum conservation, one infers that because the muons in the above experiment are exclusively left-handed, the accompanying neutrinos are exclusively left-handed.

handed muons never exist in nature? Clearly this is not so, for we have noted that a particle with *finite* rest mass, e.g., a muon, need not always be left-handed. Merely by jumping on a reference frame that moves faster than our left-handed muon, we can convert it into a right-handed one. Left- or right-handedness clearly is not an intrinsic property of a muon. But the same reasoning does not apply to a neutrino, which with its zero rest mass always moves at the speed c. As a result we never can find a reference frame to jump onto in order to pass a neutrino. Left-handedness is indeed an intrinsic property of a neutrino; once left-handed, always left-handed. Therefore, the exclusive appearance of left-handed neutrinos in the π^+ decay suggests that perhaps all neutrinos in nature are left-handed. Indeed, many experiments involving a variety of interactions have now established that all neutrinos, both muon and electron types, are left-handed, and that all antineutrinos are right-handed. Establishing this fact is equivalent to establishing the nonconservation of parity.

Neutrinos, left-handed; antineutrinos, right-handed

The overthrow of parity conservation is like a warning flag, a reminder that physical theory merely represents the status quo, not final dogma. Physics is an open and evolving adventure. The Newtonian theory, once thought so conclusive, has been assimilated into new twentieth-century structures; who knows how soon these structures will be enfolded into a deeper, simpler strategy?

SUMMARY

Unless some conservation law precludes a process, the process will occur. This, however, is not to say that the probability of occurrence is the same for all allowed events. To account for events that are observed and others that are never observed the following additional conservation laws are forced upon us: electron number, baryon number, and muon number. For every particle there exists a corresponding antiparticle.

The four fundamental interactions are strong, electromagnetic, weak, and gravitational. The family-conservation laws, as well as those for momentum, mass-energy, electric charge, and angular momentum, are universal: they hold in all four interactions. Some additional conservation laws hold in some interactions but not all. For example, parity conservation applies to all interactions except the weak.

CHAPTER 28 ELEMENTARY-PARTICLE PHYSICS

PROBLEMS

28-1 A positron and an electron, each with a kinetic energy of 1.02 MeV, collide head on and produce two photons. (*a*) What is the energy of each photon? (*b*) Suppose that these photons each produce electron-positron pairs. What is the kinetic energy of each electron or positron?

28-2 If photons with an energy greater than 1.02 MeV impinge upon absorbing material, one always observes some photons with an energy of exactly 0.51 MeV emerging from the material. Give the physical processes responsible for the 0.51-MeV photons.

28-3 An electron collides with a second free electron initially at rest, and an additional positron and electron are created in the collision. (*a*) Show that this process is not ruled out by the conservation laws. (*b*) What is the minimum kinetic energy for the electron initially in motion for which this process can take place? (*c*) Now consider two electrons in motion with the same kinetic energy colliding head on and also producing an electron-positron pair? What is the minimum kinetic energy for each of the two electrons in this situation?

28-4 (*a*) A proton captures an antineutrino of the electron family to produce a neutron and positron. What is the (approximate) minimum neutrino energy for this process to occur? (*b*) What is the (approximate) minimum energy of an antineutrino of the muon family which will produce a neutron and muon when captured by a proton?

28-5 Listed below are a number of conceivable decay processes. Some may occur; others never do. Determine whether the conjectured process is possible and if not, cite the conservation law which precludes it. Assume in each instance that the original particle is isolated and initially at rest unless its energy is given.

(*a*) $\pi^0 \to \gamma$
(*b*) $\pi^0 \to \gamma + \gamma$
(*c*) $n \to K^+ + \pi^0 + e + \bar{\nu}_e$
(*d*) $K^+ \to \pi^+ + \pi^0$
(*e*) $\Omega^- \to \Xi^- + \pi^0$
(*f*) $\gamma(5.0 \text{ GeV}) \to p + \Lambda^0 + \pi^0$
(*g*) $n \to p + e + \nu_e$
(*h*) $n \to p + \mu^- + \bar{\nu}_\mu$
(*i*) $\mu^- \to e + \nu_e + \bar{\nu}_\mu$
(*j*) $\gamma(5.0 \text{ GeV}) \to e + \bar{e}$
(*k*) $\bar{n} \to \bar{p} + \bar{e} + \nu_e$
(*l*) $K^+ \to \pi^- + \mu^+ + \bar{\nu}_\mu$

28-6 Show, by using the uncertainty principle, that the minimum rest energy of the virtual particle exchanged by a pair of nucleons interacting by the nuclear force (range $\sim 2 \times 10^{-15}$ m) is of the same order of magnitude as the pion rest energy, 10^2 to 10^3 MeV.

28-7 An electron interacts with a second electron 1 km away. Imagining the electromagnetic interaction to take place by virtue of the exchange of a virtual photon, what is the minimum energy of the photon?

28-8 It was demonstrated in (b) of Example 28-3 that an *isolated* proton at rest cannot decay into a neutron. Explain why an isolated proton with a kinetic energy even as large as ten times its rest energy cannot decay into a neutron.

28-9 In the decay $\pi^+ \to \bar{\mu} + \nu_\mu$, both $\bar{\mu}$ and ν_μ are observed to be only left-handed; moreover, in the decay $\pi^- \to \mu + \bar{\nu}_\mu$, both μ and $\bar{\nu}_\mu$ are observed to be only right-handed, thus establishing that *P is not* conserved. (a) Show that the first decay becomes the second by applying both P and C reversals simultaneously to each particle, thus showing that *PC is* conserved. (b) Show that C, therefore, *is not* conserved in the weak interaction.

28-10 An antiphysicist lives in an antiuniverse—antiprotons rather than protons, positrons rather than electrons, etc. He observes the decay of pions, having the same electric charge as the nuclei in his universe, into muons. Are these muons right- or left-handed?

ANSWERS TO SELECTED PROBLEMS

ANSWERS TO SELECTED PROBLEMS

CHAPTER 1 **1-1** (a) 10.0 m/s, (b) 32.8 ft/s, (c) 22.4 mph **1-3** (a) 2.16×10^{11} m, (b) 1.34×10^8 miles **1-5** (a) 60 mph, (b) 60 mph **1-9** 25 mph **1-11** (a) 0, (b) 100 m, (c) 0, (d) 1.89 m/s **1-13** (a) 25 mph, (b) 12.5 mph **1-15** (a) $[2\pi A_0/T_0] \cos(2\pi t/T_0)$, (b) 0, (c) $2\pi A_0/T_0$ **1-17** 2.0 m/s **1-19** 3.4 m/s

CHAPTER 2 **2-1** 2 m/s **2-3** (a) 6.0 kgm-m/s, (b) 0.2 m/s **2-5** 0.5 m/s to right **2-7** (a) 9.1 m/s, (b) no; the two bullets are not an isolated system **2-9** 1.2×10^5 cm/s **2-11** (a) 1.6 m/s to left, (b) 4.0 m/s to left **2-13** (a) 7.5 m/s, (b) 16.5 m/s, (c) 13.5 m/s **2-15** (a) 0.6 m displacement, (b) 0.4 m displacement, i.e., no change **2-17** 27 km/s

CHAPTER 3 **3-1** 20 N **3-3** (a) 107 N, (b) the ground **3-5** -12 N **3-7** (a) 160 cm/s, west, (b) 360 cm/s, west **3-9** (a) 5 N to left, (b) 5 N to left, (c) 7.5 N to left **3-11** 1.6 N upward **3-13** (a) Momentum to left at A and B, and force to right at B, (b) force to right but no momentum at C, (c) momentum to right at B and A, and force to right at B **3-15** (a) 0, (b) 0.98 N **3-17** (a) $+, 0, -, 0, +$, (b) $0, -, 0, +, 0$ **3-19** (a) 33 lb, (b) 0

CHAPTER 4 **4-1** 19.6 m/s **4-3** (a) 17 m/s, (b) 850 m **4-5** (a) 15, 10, 15, 20, 20 m/s, (b) 25, 30, 25, 20, 20 m/s, (c) 20, 20, 20, 20, 20 m/s **4-7** 80 ft/s **4-9** 12 miles **4-11** (a) 5.4 miles, (b) 4.7 min **4-13** (a) 323 N, (b) 294 N, (c) 309 N **4-15** (c) 28 N, (d) 70 N **4-17** (b) 2,170 N **4-19** (a) 66 lb, (b) 1.9 lb, (c) 10.1 lb, (d) 12 ft

CHAPTER 5 **5-1** (a) 192 miles, 28° east of north, (b) 171 miles, 20.5° east of north **5-3** $0.8v_0$ and $0.3v_0$ **5-5** (a) $y = 16$ m at 10 s, (b) $y = 24$ m at 10 s **5-7** 0.97 m/s **5-9** 0.69 kgm **5-11** (a) 64 m, (b) 32 m **5-13** 50 lb **5-15** (a) attractive, 7.3×10^{-8} N, (b) 8.9×10^{-30} N **5-17** (a) 0, (b) 2,270 lb central

ANSWERS TO SELECTED PROBLEMS

CHAPTER 6

6-1 0 **6-3** (a) 7.2 m, (b) 7.2 m, (c) 0 **6-5** (a) 4 m/s, (b) 0, (c) 6 kgm is 6 m to east, and 3 kgm is 12 m to west **6-7** (a) $\frac{4}{3}$ m/s to west, (b) 4 m/s to south, (c) 4 m/s to east, (d) 4 m/s to east **6-9** (a) 1 m/s to east, (b) 2 m/s to east and 1 m/s to west before, and both at rest after **6-11** Before collision they approach from opposite directions each with speed $v_0/2$; after collision they separate in opposite directions each with speed $v_0/2$ **6-13** Before collision they approach from opposite directions (east and west) each with speed $v_0/2$; after collision they separate in opposite directions (north and south) each with speed $v_0/2$ **6-15** (a) 30° west of north, (b) 260 mph **6-17** (a) 0.77 s, (b) 0.77 s, (c) 0.63 s, (d) as long as it takes the elevator to hit the bottom of the shaft

CHAPTER 7

7-1 3.6 kgm-m²/s **7-3** (a) 100 kgm-m²/s, (b) 112 kgm-m²/s, (c) no, (d) no **7-5** (a) 2 m from west end; remains at rest, (b) 60 kgm moves south at 1.0 m/s, and 30 kgm moves north at 2.0 m/s, (c) 30 N, (d) 30 N **7-7** (a) 18 kgm-m/s, (b) no, (c) 3.14 s, (d) man's motion unchanged **7-9** (a) 48 kgm-m²/s counterclockwise, (b) 36 N-m clockwise **7-11** About 33 **7-13** (a) 5.0 N-m, (b) 5.0 kgm-m²/s, (c) 19 rpm **7-15** Three-quarters of the way to the back end **7-17** Yes, 90 lb **7-19** (a) 50 N-m counterclockwise, (b) 20 N-m clockwise, (c) yes, (d) right

CHAPTER 8

8-1 (a) 0.5 mile, (b) 30 mph **8-3** (a) 30 m/s, (b) 30 m/s, (c) 20 m/s to west, (d) 10 m/s to east **8-7** 0.85 **8-9** (a) One ball moves out the opposite end, (b) two balls move together out the opposite end **8-11** $M/(M + m)$ **8-13** (a) Less than 90°, (b) greater than 90° **8-15** (a) +640 N-m, (b) −320 N-m, (c) +320 N-m, (d) 8.0 m/s **8-17** −15 N-m

CHAPTER 9

9-1 (a) 0.072 J, (b) 80 N/m **9-3** 13 m/s **9-5** 209 J **9-7** 53 m/s **9-9** (a) 1.7 m/s, (b) 1.0 m/s, (c) 1.4 m/s **9-11** 0.69 kgm **9-13** (a) 160 ft/s, (b) 226 ft/s, (c) 179 ft/s **9-17** 1.8 ft **9-19** (a) No; the rope under tension does work on the monkey, (b) 25.2 lb

ANSWERS TO SELECTED PROBLEMS

CHAPTER 10 10-1 546°C 10-3 230 m/s 10-5 (a) 1, (b) 1.55 10-7 (a) 3.0 atm, (b) 3.18×10^4 m/s 10-9 2.5×10^{-14} kgm

CHAPTER 11 11-1 (a) The box starts to move to the right when the marbles collide with the right side of the box; later when the left side of the box and the marbles collide, the box comes to rest and the marbles resume their original motion, (b) Nmv_0, (c) $\frac{1}{2}Nmv_0^2$, (d) although the marbles are in motion, the box remains at rest except for small random displacements about its average position, (e) 0, (f) $\frac{1}{2}Nmv_0^2$ 11-3 (a) +, (b) 0, (c) −, (d) − 11-5 (a) No, (b) no 11-7 0.05°C, (b) 2×10^5 W 11-9 $\frac{1}{10}$ raised to the power 3×10^{20}

CHAPTER 12 12-1 $\frac{1}{3}$ 12-3 1.53 AU 12-5 3.84×10^7 m, or 23,900 miles 12-7 1.3×10^2 m/s 12-9 $\frac{1}{2}$ (a) 0, (b) F_0 12-13 (a) 0, (b) 0.098 N, (c) 4.9 N 12-15 (a) 0, (b) $GmM/2R$, (c) GmM/R 12-17 $\frac{1}{2}\sqrt{Gm/r}$

CHAPTER 13 13-1 5.0×10^{-10} C 13-3 4.9×10^5 N 13-5 5.8×10^{-7} C 13-7 (a) 0.75 cm, (b) 1.12×10^{-7} C 13-9 (a) 4.3 N straight up, (b) 1.08×10^7 N/C straight down 13-11 (a) 1.1×10^{-2} C, (b) 8.8×10^{-4} C/m² 13-13 (a) 2×10^7 N/C toward center, (b) 2×10^7 N/C toward center, (c) 0 13-15 (a) unchanged, (b) inversely proportional to distance from plane surface

CHAPTER 14 14-1 (a) 0.182 J, (b) 2.0×10^{-6} J 14-3 (a) -2.0×10^{-6} J, (b) -2.0×10^{-6} J 14-5 (a) 3.4×10^2 N/C, perpendicular to the plates directed from the positive to negative charge, (b) 1.7 V 14-7 Yes, 1.9 eV 14-9 (a) 2.2×10^4 m/s, (b) 1.20×10^{-18} J, or 7.5 eV 14-13 (a) 3.7×10^4 m/s, (b) 0, and 5.2×10^4 m/s 14-15 (a) $q_0(k_e/m_0R)^{1/2}$, (b) $2q_0(k_e/3m_0R)^{1/2}$ and $q_0(k_e/3m_0R)^{1/2}$ 14-17 (a) k_eQq/d^2, (b) 0 14-19 (a) 0; k_eq_1/r^2 toward center; $k_e(q_1 + Q_2)/r^2$ toward center, (b) force on free electrons in wire is away from center

ANSWERS TO SELECTED PROBLEMS

CHAPTER 15

15-1 1 cm/s **15-3** (a) 2 Ω, (b) 18 W **15-5** 80 A into the paper **15-7** (a) zero, (b) zero, (c) 7.2×10^{-12} N vertically downward, (d) 7.2×10^{-12} N vertically upward, (e) 4.3×10^{-12} N vertically downward, (f) 7.2×10^{-12} N eastward, (g) 7.2×10^{-12} N westward **15-9** (a) west, (b) 7.5×10^{-2} N **15-11** (a) 2.9×10^7 m/s, (b) 8.7×10^{-8} s, (c) 8.6×10^6 V **15-13** (a) west, (b) 4.3×10^{10} m/s (an impossible speed, since it exceeds that of light; more about this in Chapter 22), (c) $F_m = 2.9 \times 10^7 \, F_g$ **15-15** 0.2 N/m **15-17** (a) 1 mA, (b) 13 T **15-19** (b) Perpendicular to the plane of the circular loop

CHAPTER 16

16-1 (a) 10 A, (b) 0.05 Ω, (c) 1.0 V, (d) 15 W **16-3** (a) 3 A, (b) 4 V, (c) 2 V, (d) 0 V, (e) 0 Ω **16-5** (a) 16 V, (b) 22 V, (c) 54 V **16-7** (a) 0.016 Wb, (b) zero, (c) zero **16-9** (a) 0.06 V, counterclockwise, (b) 0.19 V, counterclockwise, (c) 0.15 V, counterclockwise **16-11** (a) 0.06 V, (b) a to b, (c) 0.72 mW **16-13** (a) positive z direction, (b) 0.04 T/s **16-17** (a) 1 Wb/s, (b) 1 V, (c) 0.1 A, (d) counterclockwise, (e) 0.10 W, (f) 0.05 N, (g) force of hand to left

CHAPTER 17

17-1 counterclockwise **17-3** 1.3 s **17-5** (a) 38°, (b) 33° **17-7** $t \sin(\theta_1 - \theta_2)/\cos \theta_2$ **17-9** 2.4 ft **17-11** (a) 56°, (b) glass **17-13** 760 cm²

CHAPTER 18

18-1 (a) halved, (b) doubled, (c) doubled **18-3** 16.5 mm to 16.5 m **18-7** (a) 0.15 mm, (b) double source frequency, halve D, double d, double refractive index (e.g., fill space between slits and screen with material such as zincite, for which $n = 2.013$) **18-9** (a) 100 cm, (b) one transmitter is oscillating with the opposite polarity relative to the other **18-11** (a) 10 m, (b) north-south **18-13** (a) Distance of a particular principal maximum from center is halved, (b) principal maxima unchanged in position but narrower, (c) separation of principal maxima is doubled

CHAPTER 19

19-1 10^3 km **19-3** 5 m **19-5** 3.3×10^{14} Hz **19-7** (a) 83 m, (b) 0.02 m, (c) distance in (a) doubled and in (b) unchanged, (d) distance in (a) reduced by factor 4 and in (b) reduced by factor 2 **19-11** (a) 0.8 cm, (b) 0.9 cm, (c) doubled, reduced by a factor $\frac{1}{2}$, (d) reduced by a factor $\frac{1}{4}$, doubled

ANSWERS TO SELECTED PROBLEMS

CHAPTER 20 20-1 North 20-3 (a) positive z direction, (b) 2.0×10^{-7} T 20-5 $B^2/8\pi k_m$ 20-7 (a) 6.7×10^{-7} N, (b) 3.3×10^{-7} N

CHAPTER 21 21-1 6.3×10^{-5} s 21-3 2.3 h 21-5 (a) $0.8c$, (b) 2.5×10^{-8} s 21-7 (a) 35.8 years, (b) 3.4 light-years 21-9 34.7 m 21-11 (a) 117 m, (b) 42 m, (c) no 21-13 (a) 8 and 24 ft, (b) 8 and 14 ft 21-15 $0.98c$ 21-17 $0.995c$

CHAPTER 22 22-1 (a) 9.1×10^{-31} kgm, (b) 9.1×10^{-31} kgm, (c) 45.5×10^{-31} kgm 22-3 $\tfrac{1}{2}L_0$ 22-5 $0.87c$ 22-7 (a) 1.56 GeV, (b) 0.62 GeV 22-9 (a) 1 GeV/c, (b) 1.7 GeV/c, (c) 1 GeV/c 22-11 (a) 2.7×10^{-30} kgm, 1.0 MeV, (b) 2.3×10^{-27} kgm, 1 GeV, (c) 5.3×10^{-26} kgm, 29 GeV 22-13 (a) 5×10^7 kgm, (b) no 22-17 (a) 1.0 MeV, (b) 1.4 MeV/c, (c) 4.7×10^{-3} T

CHAPTER 23 23-1 1.48 eV 23-3 9.9 eV 23-5 7.2×10^{-34} J-s 23-7 (a) 6.2×10^{-9} eV, (b) 4.1×10^{-7} eV, (c) 1.24×10^{-5} eV, (d) 2.48 eV 23-9 (a) 16×10^8, (b) 6.0×10^{24}, (c) 3.0 MW 23-11 (a) 0.050 Å, (b) 0.052 Å, (c) 0.002 Å

CHAPTER 24 24-1 (a) 7.7 MeV, (b) 3.0×10^{-14} m, (c) 7.7 MeV, (d) 1.5×10^{-14} m 24-3 (a) 1.2×10^3 Å, (b) 9.1×10^2 Å at the ionization limit 24-7 Three 24-9 (a) 54.3 eV, (b) 0.264 Å

CHAPTER 25 25-1 6.6×10^{-34} m 25-3 (a) 3.8×10^{-2} eV, (b) 1.5 Å 25-5 (a) 0.12Å, (b) 1.7° 25-9 (a) $\gtrsim 10^{-28}$ m/s $= 10^{-18}$ Å/s, (b) $\gtrsim 10^7$ m/s 25-11 16

CHAPTER 26 26-1 $\tfrac{1}{6}, \tfrac{1}{3}, \tfrac{1}{2}, \tfrac{2}{3}, \tfrac{5}{6}$ Hz 26-5 (n,l,m_l,m_s): $(0,0,0,\pm\tfrac{1}{2})$, $(1,0,0,\pm\tfrac{1}{2})$, $(2,1,0,\pm\tfrac{1}{2})$ $(2,1,1,\pm\tfrac{1}{2}), (2,1,-1,\pm\tfrac{1}{2})$ 26-9 (a) $y = 2A \cos(n\pi vt/L) \sin(n\pi x/L)$, where $n = 1,2,3,$ and 4, (b) $nv/2L$, $2L/n$

ANSWERS TO SELECTED PROBLEMS

CHAPTER 27

27-7 $^{17}_{8}O$ **27-9** $\frac{1}{16}$ **27-11** $1/e$

CHAPTER 28

28-1 (a) 1.53 MeV, (b) 0.26 MeV   **28-3** (b) 3.06 MeV, (c) 0.51 MeV
28-5 (a) No, momentum, (b) yes, (c) no, baryon, (d) yes, (e) yes, (f) no, charge, (g) no, electron number, (h) no, energy, (i) no, electron number, (j) yes, (k) yes, (l) no, electric charge   **28-7** $\sim 10^{-9}$ eV

INDEX

INDEX

A (see Ampere)
Å (see Angstrom)
Abbreviations of prefixes, table, 20n.
Absolute temperature, 207
Absolute zero, 207
Absorption of electromagnetic waves, 454–458
Acceleration:
 average, 48
 of center of mass, 118–119
 central, 105
 centripetal, 105
 constant, 64–65
 of gravity, 52–55
 instantaneous, 48
 radial, 105
 transformation, 129
Accelerator:
 cyclotron, 328
 electron linear, 653
 high-energy, 668
 Van de Graaff, 303
Acoustic spectrum, 415
Action:
 at a distance, 260
 and reaction, 71
Activity, 638
Addition:
 algebraic, 45
 vector, 85
Adiabatic process, 232
Affinity, electron, 613
Algebraic sum, 45
Alkali metal, 615
Alpha particle, 548
 scattering, 549
Ampère, A. M., 311, 337
Ampere (unit), 314, 333
Amplitude, 410
Angle:
 critical, 388
 of incidence, 377

Angle:
 of reflection, 377
 of transmission, 384
Angstrom (unit), 414
Angular momentum, 132–141
 conservation, 141, 147–149
 orbital, 139
 of particle, 134–141
 of planet, 241
 quantization, 562–565, 605–608
 spin, 138, 139, 679
 and torque, 144–146
Angular speed, 139
Annihilation, pair, 655
Antineutrino, 667
 capture, 670
Antinode, 595
Antiparticle, 655
Antiproton production, 665
Apogee, 262
Astronomical constants, table, 261
Astronomical unit, 240
Atom, nuclear model of, 548–549
Atomic clock, 20
Atomic model:
 Bohr, 552–556
 classical planetary, 549–552
 magnetic material, 337
Atomic nucleus, 548
Atomic number, 624
Atomic standards:
 of length, 20
 of time, 20
Atoms in groups, 613–620
AU (see Astronomical unit)
Avogadro, A., 214
Avogadro's hypothesis, 213
Avogadro's number, 218
Azimuthal angle, 599

Balmer, J., 551
Balmer series, 551

Barrier penetration, 602
Baryon, 662
Battery, 351
Beam propagation, 428–432
Becker, H., 625
Becquerel, H., 548
Bernoulli, D., 208, 209
Beta decay, 636, 667–671
Beta particle, 548
BeV, 522n.
Binary stellar system, 262
Binding energy:
 hydrogen atom, 561
 nuclear, 630–632
Bohr, N., 552, 558, 586
Bohr model of hydrogen atom, 552–556
Boltzmann, L., 208
Boltzmann constant, 214
Bonding:
 covalent, 616
 ionic, 613
Born, M., 588n.
Bothe, W., 625
Boyle, R., 206
Bragg, W. L., 571
Bragg planes, 571
Bragg reflection, 572
Bragg relation, 572
Brahe, T., 239
Brownian motion, 208, 218, 240n.
Bubble chamber, 324

c, 381
c (calculus symbol), 11n.
C (see Coulomb)
Cal (see Calorie)
Calorie, 226
Capacitance, 305–306
Capacitor, 367
Cavendish, H., 249

INDEX

Cavendish experiment, 249–251
Celsius temperature, 206
Center of gravity, 155
Center of mass, 114–116
   acceleration of, 118–119
   reference frame, 125, 126
   velocity of, 114
Centigrade temperature, 206n.
Central acceleration, 105
Central force, 104, 145, 147, 240
Centripetal acceleration, 105
CERN, 668
Chadwick, J., 625
Chamberlain, O., 663
Charge, 236
   electric, 266–269
   gravitational, 238
   magnetic, 236
   reversal of electric, 683
Charles, J., 206
Circular motion, uniform, 103–105, 112
Clausius, R. J. E., 208
Cockcroft, J. D., 648
Coefficient of restitution, 163–166
Colliding beams, 667
Collisions:
   in center-of-mass reference frame, 125–126
   glancing, 82–83
   head-on, 15
   in mass, definition, 16
Color, 414–415
Complementarity principle, 585–586
Components, vector, 87–88
Compressibility, 204
Compton, A. H., 537
Compton effect, 537–541
Compton relation, 540
Compton wavelength, 540
Condensed states of matter, 617

Conductor, 297–300
Conservation law(s), 26
   of angular momentum, 141, 147–149
   of baryon number, 662–665
   of electric charge, 270
   of electron number, 659–662
   of energy, 182
   of mass, 26–27
   of momentum, 28–30, 83–85, 89
   of muon number, 671–674
   universal, 652–655
Conservative interaction, 183
Constant acceleration, 64–65
Constants:
   astronomical, table, 261
   physical, table (*see* inside back cover)
Constructive interference, 406
Conversion factors:
   commonly required, table (*see* inside back cover)
   for energy, 171
Corner reflector, 394
Correspondence principle, 558–561, 564
Cosmic radiation, 326
Coulomb, C., 273
Coulomb (unit), 267, 333
Coulomb's law, 271–273
Covalent bonding, 616
cps (unit), 414
Critical angle, 388
Critical point, 619
Current:
   electric, 313
   induced, 346
   in loop, 336
   magnetic force on conductor, 329–331
Cyclotron, 328
Cyclotron frequency, 325

Dalton law of partial pressures, 217
Davisson, C. J., 575
de Broglie, L., 568
de Broglie wave speed, 569
de Broglie wavelength, 568
Decay:
   alpha, 635
   beta, 636, 667–671
   constant, 639
   electron capture, 637, 670
   gamma, 634
   law, 638–643
   neutron, 667
   pion, 684
   positron, 637
δ, 253
Δ, 8
Destructive interference, 406
Deuterium, 631
Deuteron, 631
Dicke, R. H., 56, 238
Diffraction, 428–444
   electron, 575
   Fraunhofer, 437, 440–441
   Fresnel, 437–439
   single-slit, 437, 576
   x-ray, 570
Dilation of time, 474–479
Dilute gas, 204–206
   kinetic model, 208–210
Dipole, electric, 288
Displacement, 8
Distances, table, 7n.
Double-slit experiment, 391–393, 411–414
Dyn (*see* Dyne)
Dynamics, relativistic, 506–524
Dyne (unit), 42

e, 268, 165
Earth satellite, 106

East-west effect, 327
Einstein, A., 56, 240n., 470, 487, 507, 532
Elastic object, 166
Electric charge, 266–269
  conservation, 270
  quantization, 271
Electric current, 313
Electric dipole, 288
Electric field, 275–277
  changing, 346
  of continuous charge distribution, 279–281
  energy of, 284–285
  energy density of, 285
  induced, 357
  lines of, 278
Electric flux, 367
Electric interaction, 266–304
  Coulomb's law, 271–273
Electric motor, 337
Electric potential, 290–292
Electric potential energy, 294–295
Electrolyte, 298
Electromagnetic force and photon exchange, 675
Electromagnetic rocket, 463
Electromagnetic spectrum, 452, 455
Electromagnetic waves, 448–465
  absorption of, 454–458
  emission of, 454–458
  momentum of, 463–464
  quantization of, 532
  sinusoidal, 453
Electromagnetism and relativity, 341
Electromotance, 347–350
  in moving conductors, 361–364
Electromotive force, 350n.
Electron accelerator, 653
Electron affinity, 613
Electron capture, 637, 670
Electron charge, 268

Electron charge-to-mass ratio, 342
Electron diffraction, 575
Electron mass, 270
Electron number conservation, 659–662
Electron-positron pair, 656
Electron spin, 338, 607
Electron volt (unit), 292, 521
Electroscope, 302–303
Electrostatic equilibrium, 298
Electrostatic shielding, 299
Elementary particles, 652
  properties, table, 654
Emf (see Electromotive force)
Emission of electromagnetic waves, 454–458
Energy:
  binding, 561, 630–632
  conservation, 182–184
  density: of electric field, 285
    of magnetic field, 586
  of electric interaction, 290
  internal, 225
  ionization, 555
  kinetic, 168, 210
  potential, 182
    (See also Potential energy)
  quantization, 548, 596
  relativistic, 512–515
  rest, 513
  table of approximate values, 173n.
  thermal, 220
  total, 184
    relativistic, 513
Energy-level diagram, 554
Energy-work relationship, 168–170
  variable force, 174
English system of units, 20
Eötvös, R., 55
Equilibrium, 46, 102–103
  electrostatic, 298
  rotational, 149–150
  static, 46

Equilibrium:
  thermal, 222
Equipotential surface, 296
Erg (unit), 170
Escape velocity, 256
Ether, 469, 497
eV (see Electron volt)
Exchange of velocities, 158, 159
Excited state, 556
Exclusion principle, 610
Expansibility, 205
External force, 118

Fahrenheit temperature, 207
Faraday, M., 279, 299, 347
  law of, 357
Faraday effect, 347
Faraday ice-pail experiment, 299
Field:
  electric, 275–277
  gravitational, 260–261
  magnetic, 319–321
  radiation, 459
First law of motion, 70
First law of thermodynamics, 225–227
Fission, nuclear, 645
Fizeau, A. H. L., 383
Flux:
  electric, 367
  magnetic, 353–356
Foot-pound, 170–171
Force, 40–41
  and acceleration, 94–97
  average, 41
  central, 104, 240
  constant, 73
  external, 118
  in fundamental interactions, 674–679
  gravitational, 236, 678
  instantaneous, 41
  internal, 119

Force:
   isotropic, 205
   magnetic, 312, 496
     right-hand rule, 322
   net, 45
   nonconservative, 193–194, 348
   normal, 107
   nuclear, 629–630, 677
   radiation, 460–461
   reaction, 71
   of spring, 72–73
   superposition principle, 40–45, 98–100
   vector, 94–95
   velocity-dependent, 312
Frame of reference, 120–121
   center-of-mass, 125, 126
   inertial, 123–124
Franck, J., 557
Fraunhofer, J. von, 437n.
Fraunhofer diffraction, 437, 440–441
Fresnel, A., 430
Fresnel diffraction, 437, 440–441
Fresnel zone, 430
Frequency, 410
   cyclotron, 325
Frictionless glider, 4–5
Friedrich, W., 570
Fringes, 412
Ft-lb (*see* Foot-pound)
Foucault, J., 390
Foucault pendulum, 123

g, 53
G (*see* Gauss)
G, 243
Galilean velocity transformation, 121
Galilei, G., 7, 55, 70, 98, 121, 238, 380
Galvanometer, 343
Gamma decay, 634

Gamma ray, 548
Gas:
   dilute, 204–206
   ideal, 210–212
   law of perfect, 206–207
Gauss (unit), 320
Gay-Lussac, J., 206
Gedanken experiment, 133n.
Geiger, H., 549
General theory of relativity, 56
Germer, L. H., 575
GeV, 522
Gilbert, W., 326
Glancing collision, 82–83
Glider, frictionless, 4–5
Grating spectrometer, 416–419
   for x-rays, 570
Gravitational acceleration, 52–55
Gravitational charge, 237–239
Gravitational constant, 249–251
Gravitational field, 260–261
Gravitational force, 236
   and gravitons, 678
Gravitational interaction, 236
   between spherical objects, 246–247, 262
Gravitational mass, 238
Gravitational potential energy, 189, 252–257
Gravitational waves, 261, 678
Graviton, 678
Grimaldi, F. M., 437
Ground, electric, 301
Ground state, 555

h, 529, 532
$\hbar$, 606
Half-life, 641
Hall, E. H., 330
Hall effect, 330
Halogen, 615
Head-on collision, 15

Heat, 224
Heisenberg, W., 675
Heisenberg uncertainty principle, 575–578, 675–676
Helmholtz coil, 335
Henry, J., 357n.
Hertz, G., 557
Hertz, H. R., 367, 528
Hertz (unit), 414
Homogeneity of space, 14
Hooke, R., 73
Hooke's law, 73
Huygens, C., 159, 166, 381
Huygens' principle, 432–433
   reflection, 433–435
   refraction, 434–437
   and Snell's law, 435
Huygens' wavelet, 435
Hydrogen atom:
   Bohr model, 552–556
     energy, 555
     radius, 555
   lines, 551
   radius, 561
   21-cm line, 609
   wave theory, 603
Hydrogen molecule, 616
Hz (*see* Hertz)

Ice-pail experiment, 299
Ideal-gas law, 210–212
Impulse, 42
Incident angle, 377
Incident ray, 377
Index of refraction, 384
Induced current, 346–347
Induced electric field, 357
Induced electromotance, moving conductors, 361–364
Induced magnetic field, 366
Induction, electromagnetic, 347
Induction law:
   of Faraday, 357–359

Induction law:
   of Maxwell, 366–368
Inelastic collision, quantum theory, 557
Inertia:
   law of, 6–8, 14
   moment of, 139
Inertial reference frame, 36, 123
Infrared radiation, 454
Instantaneous velocity, 11
Insulator, 297
Intensity, 380
Interaction:
   classical, 236–237
   conservative, 183
   electric, 266–304
   fundamental, 236–261
   gravitational, 236
   strong, 237, 629
   weak, 237, 678
Internal energy, 225
Internal force, 119
Internal reflection, 388
Internal resistance, 352
Interference, 406
Interference fringe, 412
Invariance of momentum conservation, 124–125
Inverse-square law:
   electric force, 272
   gravitational force, 242, 247–249
   for intensity, 380
   magnetic force, 313
Ionic bonding, 613
Ionization energy, 555
Isolation, 3–6, 15
Isotropic force, 205
Isotropy of space, 14

J (*see* Joule)
Joliot-Curie, F., 626
Joliot-Curie, I., 626
Joule (unit), 170, 171

k, 214
k_e, 273
k_m, 313
Kelvin, W. T., 208
Kelvin temperature, 207
Kepler, J., 70, 239
Kepler's laws, 239–242
keV, 522
kgm (*see* Kilogram)
Kilogram (unit), 19
Kinetic energy, 168, 210
   retrieval principle, 168
   and temperature, 215–216
Kinetic model, dilute gas, 208–210
Knipping, P., 570
Kronig, R. deL., 208

Laser, 376*n*., 446, 500
Laue, M. von, 570
Law in physics, 8
   universal, 652–655
Lawrence, E. O., 329
Lebedev, P., 379
Lee, T. D., 683*n*.
Left-handed object, 681
   neutrino, 685
Leibnitz, G. W. von, 166
Lenard, P., 529
Length, proper, 484
Length contraction, 480–484, 496
Lenz, H. F. E., 358
Lenz' law, 357
Lever arm, 145
Lifetime of particle, 480
Light:
   as electromagnetic effect, 448
   familiar observations, 376–377

Light:
   momentum of, 460–464
   particle theory, 376–390
   pipe, 395
   pressure, 379
   speed of (*see* Speed, of light)
   wave theory, 398–424
Light-year, 434*n*.
Lines of electric field, 278
Livingston, M. S., 329
Lyman, T., 551

m (*see* Meter)
Magnet, 310, 337
   atomic model, 337
Magnetic bottle, 327, 647
Magnetic charge, 310
Magnetic constant, 313
Magnetic effects:
   in relativity, 341
   summary of, 311
Magnetic energy density, 586
Magnetic field, 319–321
   changing, 346
   of current loop, 333–336
   of earth, 326
   induced, 366
   lines, 320
   particles in, 323–325
   and relativity, 495–496
Magnetic flux, 353–356
Magnetic force, 312, 322, 496
   electromagnetic wave, 460
   and reference frames, 339–341, 495–497
Magnetic interaction, 310–341
Magnetic pole, 310
Magnetism:
   induced current, 346
   terrestial, 326
Marsden, E., 549
Maser, 501

Mass, 15–19
  atomic, table, 648, 649
  center of, 114–116
  conservation law, 26–27
  gravitational, 238
  inertial, 238
  proper, 509
  relativistic, 509
  rest, 508, 516
    zero, 509
  unit (atomic), 521
  and weight, 55–56
Masses, approximate values, table, 20n.
Matter waves, 568–570, 575
  probability interpretation, 586–588
Maxwell, J. C., 208, 279, 366, 448, 469
Maxwell's law, 366–368, 448
Mean life, 642, 648
Mean square speed, 212
Melville, T., 551
Meter (unit), 19
Metric system, 19
  abbreviations, table, 20n.
MeV, 522
Michelson, A. A., 469, 498
Michelson interferometer, 498
Michelson-Morley experiment, 469, 497–501
Mirror, 377
Mks unit system, 19
Mole, 218n.
Molecular potential energy, 218
Molecular speed, 212
Moment of inertia, 139
Moment arm, 136
Momentum, 27
  angular, 132–141
  conservation law, 28–30, 83–85, 89
  and reference frames, 35

Momentum:
  of light, 460, 463–464
  of photon, 536
  relativistic, 509, 511–512
Morley, E. W., 469, 498
Motor, 337
Moving conductors, induced electromotance, 361–364
Mu-mesic atom, 561
Muon, 561, 671
  family, 673
  number, 671

N (see Newton)
Net force, 45
Neutrino, 637, 667, 684–685
Neutron:
  discovery, 625–629
  properties, 629
  thermal, 589
Newton, I., 42, 55, 70, 106, 240, 246, 389, 390, 420, 469
Newton (unit), 42
Newton's laws of motion, 70–72, 119
Newton's rings, 390–391, 420–422
Node, 595
Nonconservation of parity, 679–685
Nonconservative force, 193–194, 348
Nonisolated system, 193
Normal force, 107
Nuclear atom, 548–549
Nuclear binding energy, 630–634
Nuclear fission, 645
Nuclear force, 629–630, 677
Nuclear fusion, 645
Nuclear physics, 624–647
Nuclear reactions, 644–647
Nuclear stability, 633
Nucleon, 630

Nucleus, 548
  daughter, 634
  electron-proton model, 624–625

Oersted, H. C., 310
Oersted effect, 310
Ohm, G., 315
Ohm's law, 315, 351
Ω (see Ohm's law)
Orbital angular momentum, 139
Orbital orientation quantum number, 609
Orbital quantum number, 606
Order-disorder, 220–222, 229–231

Pair annihilation, 655
Pair production, 655
Parity, 683
  nonconservation, 679–685
Particle, 2
  elementary, table, 654
  of light, 378
  properties of waves, 528–543
  theory of refraction, 388–390
  wave properties of, 568–589
  of zero rest mass, 518
Paschen, F., 551
Pauli, W., 610, 637
Pauli exclusion principle, 610
Pendulum, Foucault, 123
Perfect-gas law, 206–207
Perigee, 262
Period, 107, 139, 410
Periodic motion, 410
Periodic table, 610–612
Periodic waves, 410
Perrin, J., 208
Phase diagram, 619
Photoelectric effect, 528–535
  classical theory, 530–531
  quantum theory, 532–535

Photon, 26n., 532
  energy, 532
  exchange in electromagnetic force, 675
  momentum, 536
  rocket, 519
  virtual, 676
Pion, 677
  decay of, 684
Planck, M., 532
Planck's constant, 532
Planetary (classical) model of atom, 549–552
Planetary motion, Kepler's laws, 239–242
+ (symbol for extra material), 14n.
Pole, magnetic, 310
Positron, 655
Positron decay, 669
Potential, electric, 290–292
Potential energy, 182
  barrier, 259
  electric, 294–295
  gravitational, 189–191, 252–257
  of spring, 185–187
Pound-feet, 170
Power, 348
Pressure, 205
  of light, 379
Probability in radioactive decay, 641
Probability density, 600
Probability interpretation, matter waves, 586–588
Problem answers, 689
Production, pair, 655
Propagation of beam, 428–432
Proper length, 484
Proper mass, 509
Proper time, 478
Proton mass, 305
Proton properties, 629

Quantization:
  of angular momentum, 562–565, 605–608
  of electromagnetic waves, 532
  of energy, 596
    atomic systems, 548–565
Quantum hypothesis, 532
Quantum jump, 556
Quantum number:
  hydrogen atom, 607
  orbital, 605
  orbital orientation, 609
  principal, 603
  spin orientation, 608
Quantum systems of particles, 592–621
Quantum theory, 528
Quantum transition, 556
Quark, 271

Radial acceleration, 105
Radiation:
  from accelerated charge, 457
  cosmic, 326
  field, 459
  force, 460
  pressure, 379, 461
Radio astronomy, 609
Radioactive decay, 634–638
  law of, 638–643
Radioactivity, natural, 548
Range of trajectory, 110
Rare-gas atom, 614
Ray of light, 377
Reaction, nuclear, 645
Reaction force, 71
Reference frame, 120
  center-of-mass, 125, 126
  inertial, 123
  and magnetic force, 339–341, 495–497

Reference frame:
  and momentum conservation, 35–36
  speed of light, 468–474
Reflected ray, 377
Reflection:
  of light, 377
  total internal, 388
  of waves, 407
Refraction of light, 384
  particle theory, 388–390
Refraction index, table, 384
Relative speed, 160
Relativistic dynamics, 506–524
  computations, 521–522
  units, 521–522
Relativistic energy, 512–515
Relativistic mass, 509
Relativistic momentum, 509, 511–512
Relativistic velocity transformation, 489–494
Relativity:
  of electric and magnetic fields, 495–496
  general theory, 56
  magnetic effects, 341
  principle of, 507
  of simultaneity, 487–489
Resistance, 313, 351
  internal, 352
Resistor(s), 352
  in parallel, 352
  in series, 353n.
Resonance, 593
Rest energy, 513
Rest mass, 508, 516–517
  particle with zero, 518–519
Restitution coefficient, 163–166
Retrieval principle, 161
  for kinetic energy, 168
  for relative speed, 161
  for *vis viva*, 166–167

Right-hand rule:
   for magnetic fields, 320
   for magnetic forces, 322
Right-handed object, 681
   antineutrino, 685
Rms (*see* Root-mean-square speed)
Rocket, 33
   electromagnetic, 463
   photon, 519
Roemer, O., 381
Root-mean-square speed, 212
Rotational equilibrium, 149, 150
   static, 149
Rowland, H. A., 310
Rutherford, E., 548, 552
Rydberg, J. R., 551
Rydberg constant, 551

s (*see* Second)
Salt molecule, 613
Satellite, 106
   "fixed," 245
Scattering:
   of alpha particles, 549
   of photons, 538
Schrödinger, E., 592
Schrödinger equation, 592
Scientific notation, 7n., 19n.
Second (unit), 20
Second law:
   of motion, 70–71
   of thermodynamics, 228–231
Segre, E., 663
Shell, atomic electron, 603, 607, 611
Σ, 29
Simple harmonic motion, 112, 180, 200
Simultaneity, 487–489
Single-slit diffraction, 437–441
   and uncertainty principle, 576–578

Sinusoidal motion, 112, 180, 200
Sinusoidal wave, 410
Snell, W., 384
Snell's law, 383–385, 435
Sodium spectrum, 419
Solenoid, 335
Space, uniformity of, 14–15
Space pistol, 32
Spectrometer, 416
Spectrum:
   absorption, 556
   acoustic, 415
   continuous, 551
   dark line, 556
   electromagnetic, 456
   emission, 551
   hydrogen, 551
   line, 551
   visible, 415
Speed, 11
   angular, 139
   of light, 340, 380–381, 452
      measurement of, 380–381
      in a medium, 435
      and reference frames, 470–474
      universal constancy of, 468–470
      through water, 390
   molecular, 212
   root-mean-square, 212
   table of approximate values, 9n.
Speed-retrieval principle, 158–160
Spin angular momentum, 138, 139, 679
   electron, 607
   orientation quantum number, 608
Spring potential energy, 185
Standing waves, 593
Static equilibrium, 46
Stationary state, 553
Storage ring, 668

Strong interaction, 237, 629
Subtraction, vector, 94
Superconductivity, 618
Superfluid, 619
Superposition:
   of forces, 44, 98–100
   of waves, 405–407
Symmetry, 16n.
Systems of particles, quantum theory, 592–621

T (*see* Tesla)
TCP reversal, 682
Temperature, 223
   absolute, 207
      zero of, 207
   Celsius, 206
   centigrade, 206n.
   Fahrenheit, 207
   Kelvin, 207
   and kinetic energy, 215–216
   mechanical interpretation of, 216
   and thermal equilibrium, 222–224
Tension, 74n.
Tesla (unit), 320
Thermal energy, 220–222
Thermal equilibrium, 222–224
Thermal neutron, 589
Thermal oscillations, 220, 222
Thermodynamics:
   first law of, 225–227
   second law of, 228–231
Thermometer, 223
Third law of motion, 70–71
Third law of Newton, 119
Thomson, J. J., 342, 528, 548
Thomson, G. P., 575
Thought experiment, 133
Threshold frequency, 529

INDEX

Time:
- dilation, 474–479
- interval, table, 8n.
- proper, 478
- reversal, 661, 682
- standard unit of, 20

Torque, 144
- external, 148
- internal, 148
- magnetic, 336

Torsion balance, 274
Torsional stiffness, 249
Transformation, velocity: classical, 470
- relativistic, 489

Transition element, 612
Trigonometric functions, illustration (*see* inside back cover)
Triple line, 619
Type style (font):
- boldface, 85
- sans serif, 27n.

u (*see* Mass unit)
Uncertainty principle, 575–578, 675–676
Unified atomic mass scale, 624
Uniform circular motion, 103–105, 112
- work done in, 175

Uniformity of space, 14–15
Units, 19–20
- English system, 20
- metric system, 19
- in relativistic dynamics, 521–522

V (*see* Volt)
Valence, chemical, 271, 614, 622
Van Allen, J. A., 327
Van Allen belts, 327

Van de Graaff, R. J., 303
Van de Graaff generator, 303–304
Vector, 85
- addition, 85
- by component method, 87–88
- components, 87–88
- subtraction, 94

Velocity:
- average, 9
- of center of mass, 114, 115
- components, 91
- escape, 256
- instantaneous, 11
- of light (*see* Speed, of light)
- selector, 328
- transformation: classical, 121
- relativistic, 489

Virtual photon, 676
Vis viva, 166
Visible spectrum, 415
Vision, 414
Volt (unit), 291

W (*see* Watt)
W particle, 678
Walton, E. T. S., 648
Watt (unit), 348
Wave:
- on circular membrane, 596–600
- electromagnetic, 448–465
- equation, 592
- function, 588
- gravitational, 261, 678
- longitudinal, 404
- particle properties of, 528–543
- and particle theories of light, 422–423
- periodic, 410
- propagation, coupled masses, 398–404
- properties, of particles, 568–589

Wave:
- reflection, 407–409, 433
- refraction, 433
- speed: of coupled masses, 400–404
- of light (*see* Speed, of light)
- standing, 593
- along string, 592–596
- superposition, 405–407
- theory, hydrogen atom, 603–610
- transverse, 404

Wave function, 588
Wavefront, 432
Wavelength, 410
- of light, 415

Wb (*see* Weber)
Weak interaction, 237, 678
- nonconservation of parity, 679–685

Weber (unit), 355
Weight, 55
White, E., 32
Wiegand, C. E., 663
Work, 168–170
- versus heat, 224
- by variable force, 173–174

Work-energy relationship, 168–170
- for variable force, 174

Work function, 533

X-ray, 454, 537
- diffraction, 570
- wavelength measurement, 570–572

Yang, C. N., 683n.
Young, T., 390
Young double-slit experiment, 391–393, 411–414
Ypsilantis, T., 663
Yukawa, H., 677n.